Field Marshal HRH Prince Philip, Duke of Edinburgh, KG KT OM GBE, Colonel, Grenadier Guards

BUCKINGHAM PALACE.

We all remember the formative experiences in our lives. Some look back on them with affection and nostalgia, others would be happy to wipe them from their memories completely. The Guards Depot at Caterham-on-the-Hill provided a formidable experience for many hundreds of thousands of young men on their way to join their Regiments of the Brigade of Guards.

Caterham had to prepare young men for a wide variety of challenges. Anything from conventional infantry duties to armoured units, parachutists, the Long Range Desert Group and the Special Air Service Regiment, and it did it with great success. A training depot needs to do much more than develop discipline, fitness and basic military skills. It must also instil a sense of confidence and pride, which are essential for the morale and fighting spirit of every military unit. Whatever their personal views about the Guards Depot may have been, the proof of its methods is reflected in the outstanding military reputation, which the Guards Regiments achieved in small conflicts and world wars and in the precise discipline of their ceremonial duties.

Caterham has now closed, but it would be wrong for its contribution to be forgotten. I hope that this account of the life of this famous establishment will give present and future generations a flavour of its unique character and a glimpse of the personalities who guided its activities over the years.

Philip

For the generations of Bill Browns, Lilywhites, Jocks, Micks and Taffs who survived the agonies of becoming Guardsmen at Caterham, and who have prized the achievement throughout their lives ...

THE FIVE REGIMENTS OF FOOT GUARDS

Coldstream Guards

Scots Guards

Grenadier Guards

Irish Guards

Welsh Guards

The Guards and Caterham

The Soldiers' Story

Editor: Ronald Melvin

Editor's Note

For more than a century, the Guards' Depot at Caterham-on-the Hill in Surrey played a significant part in this country's military history and, consequently, in the safeguarding of Britain's freedoms and interests. When it was built, the Boer War was still a couple of decades in the future, the two World Wars were nightmares yet to come, few in these islands had even heard of Korea, and the Irish Troubles, world terrorism, the Falklands War and Desert Storm were over an almost infinite number of future horizons.

Over the several centuries of their history, the Guards have built their reputation as fighting regiments, earning the respect of those who have opposed them in wartime. The Caterham Depot maintained the historical standards, producing a resourceful fighting man of exceptional stamina, capable of sustained individual initiative. Men who fought in Desert Storm and who have served in Northern Ireland have displayed the same qualities which turned the tide in two world wars.

The closure of the old, grey, Caterham-on-the-Hill Depot has also closed a distinctive military chapter, the ghosts of which deserve to be remembered with respect.

This is an informal and incomplete history, if it can be called that. There are gaps we have failed to fill and many significant personalities have not found their way into these pages. For that, I take responsibility and apologise. But it has one peerless quality; everything here is drawn from the recollections and researches of men from all parts of the United Kingdom who began their soldiering, willingly or reluctantly, at Caterham. Former Guardsmen of all ranks in the five regiments will recognise characters who have left indelible impressions behind - larger than life personalities who did much to convert callow and sometimes wayward youth to disciplined confidence.

Readers who never served in the Guards, and have little idea what went on behind the Depot's grim brick walls for more than a century, may find some of the terms used and the events and incidents they describe a bit strange. But they will surely recognise the continuous strand of clear, uncomplicated pride which runs through all the contributions.

It is the mark of something valuable.

The Brigade of Guards Star

Quinque Juncta In Uno – Five Joined in One

The Brigade of Guards Star was designed to represent and describe the brotherhood of the five regiments of Foot Guards: Grenadiers, Coldstream, Scots, Irish and Welsh. Designed last century to represent the three regiments then in being, the Star's Motto, *Tria Juncta In Uno*, was changed to *Quinque Juncta In Uno* in 1949 on the authority of the King.

At the same time, the colours used in the Star were specified: until then no fixed pattern and no details had been laid down.

More recently, the Brigade has become the Guards Division and the motto has been changed again — *Septimum Juncta In Uno* — to include the Horse Guards and Blues and Royals in the new arrangement.

The Beginning

Before Caterham

The origins of the Depot of the Guards regiments date back to 1832. In that year, Warley Barracks, Essex - which had been a recruit depot for the Honourable East India Company - was opened for the Brigade. Till then, Guards' recruits had been accommodated in a 'recruit house' at Wellington Barracks, London.

Warley, however, was a temporary measure. In 1833, recruits for the Coldstream and Scots Fusilier regiments were transferred to Mitcham Road barracks in Croydon; the Grenadiers followed in 1834. The Croydon Depot lasted for thirty years, commanded by Subalterns whose tour of duty there lasted fourteen days.

The Grenadiers moved to St John's Wood Barracks in 1854 and remained there until the Crimean War, 1856. In 1865, the Depot moved back to Warley. Finally, in 1877, came the transfer to Caterham.

The new Depot must have seemed too good to be true. It incorporated every advance in the military thinking of the day, including many positive developments in the field of health and hygiene. Up to this point of the 19th century, the barracks of Britain had been breeding grounds for regular epidemics which took a heavy toll of their occupants. And the barracks used for overseas postings were even worse.

A measure of the living conditions of the British soldier can be gauged from the fact that, even in peacetime, his life expectancy had been shorter than that of the average civilian.

The situation in the fleet was similar. Naval personnel suffered the same privation and lack of basic amenities to anything like a satisfactory standard. There, too, health standards were poor.

By the middle of the century, the consequences began to cause serious alarm in Government, the services, industry and commerce. Britain was the world's major trading nation, and trade followed the Flag. The nation's overseas interests - its possessions and markets depended on protection by the Army and Navy, and recruitment to both was falling to dangerous levels.

In its day, Caterham Depot was a major watershed — the first move in a national campaign to reverse the trend, and it had been planned meticulously by leading military and medical experts of the time.

So the arrivals at Caterham-on-the-Hill in 1877 found standards they had never experienced before. Instead of the disease-fostering, congested barrack rooms they had known, the living and sleeping accommodation was spacious by comparison, with greatly improved ventilation and light. Ceilings were high and allowed space for much larger windows. Naturally, facilities for drills and weapon training were of a high order, but so were those for sport, exercise and relaxation.

For recruits and old soldiers alike, it was a revelation.

The Historical Context

The 1870s, when the Guards' Depot was established at Caterham, was a decade of huge significance for Europe and America.

It saw the start of the second Industrial Revolution - a period of rapidly accelerating change which sent ripples into every aspect of life.

Alexander Graham Bell invented the telephone. Ferdinand de Lesseps, completed the Suez Canal. The development of the Bessemer furnace boosted the mass production of high quality steel. The internal combustion engine had arrived. So had the diesel engine, the typewriter, sewing machine and the filament lamp. It was to be a revolution based largely on steel and electricity, the new wonder power source.

Britain was at peace, but the horrors of war were still fresh. Men who had fought at Waterloo, where the Guards had again won high battle honours, were still alive - one of the last survivors lived till 1891 and lies in St Mary's Churchyard, Caterham-on-the-Hill. Wellington had called the battle 'the nearest run thing you ever saw in your life . . .'

And the events which would shatter peace again were already casting long shadows before them. Karl Marx had published Das Kapital in 1867. Otto von Bismarck had unified Germany in 1870. Alfred Krupp was building a huge armaments empire ...

The new Guards Depot was commissioned in this pivotal decade. It had cost £46,000 - then a huge sum. But it was a small price for the sustained contribution to Britain's defence which Caterham Barracks continued to make, well into the 1990s.

The Depot in the Census of 1881

The Census of England and Wales carried out on Sunday, April 3, 1881, showed that the 'population' of the Caterham Guards Depot had grown to a total of 503 in the four years since its commissioning.

Ignoring all unnecessary facts, the Census was a bald 'return of all the persons who slept or abode in this institution' on the night in question.

Its breakdown of personnel and their dependants showed:

Officers - 3

The Commandant: Lieut.-Colonel Charles Dashwood Thomas, Coldstream Guards (plus his wife and infant son)
Adjutant: Lieutenant Charles Daly, Grenadier Guards
Picquet Officer: Lieutenant Edward Brooke, Grenadier Guards

Other Classes of Persons - 498

Sergeants, Drummers, Rank & File, Pensioners, with their wives and children.

Grand Total - 503

Among the Depot staff, the Census identified NCOs and Trained Soldiers, plus Recruits (Privates, not Guardsmen) in training for three regiments — Grenadiers, Coldstream and Scots. Presumably because it was a Census of England and Wales, citizens of these countries were identified with their origin in terms of county and birthplace. Scots and Irish were ascribed simply 'Scotland' or 'Ireland'.

One comforting oddity thrown up by the Census arose from the fact that the official return form required the registrar to specify, against each named individual, whether or not they belonged in one of four categories:

1 Deaf and Dumb; 2 Blind; 3 Imbecile or Idiot; 4 Lunatic.

The return was 'Nil' for all four.

Topham Picturepoint

***RUDYARD KIPLING:** His Barrack Room Ballad championed the British Soldier - but it contained presentiments of a tragedy which had its origins at Caterham. The verses haunted his final years.*

Caterham and the Kipling Factor

IN 1876, a year before the Guards arrived to occupy their new Depot at Caterham, the population of this quiet, rural area was just 3,577 - or 2,250 when the inmates of the Metropolitan Asylum for Imbeciles were excluded from the calculation.

The fact that the Asylum was adjacent to the Depot was something which every recruit of every intake to all five regiments learned from sarcastic squad instructors almost before their new regimental numbers were dry. At some time during their basic training, all recruits were assured that they had 'come in the wrong gate'. Quite often, they were inclined to believe it.

Till the advent of the Depot, Caterham was a drowsy village whose people were almost all employed on the local farms. The arrival of the railway in 1856 had begun to open up other employment possibilities - notably in domestic service in the homes of the well-to-do, attracted by the new rail links with London.

James Thorne's Handbook to The Environs of London, published in 1876, talks of Caterham with 'many pleasant valleys, stored with wild thyme, sweet marjoram, burnell, boscage, and beeches'. It was idyllic, and many residents were not at all happy about the arrival of the Guards. The attitude was far from unusual in peacetime. And it was soon to be immortalised in verse which still strikes at the public conscience to-day:

Then it's Tommy this, an' Tommy that
an' Tommy 'ow's yer soul?'
But it's 'Thin red line of 'eroes'
when the drums begin to roll

For it's Tommy this, an' Tommy that
an' 'Chuck him out, the brute!'
But it's 'saviour of 'is country'
when the guns begin to shoot ...

Kipling wrote his Barrack Room Ballad about the swings in public attitudes to Tommy Atkins, the nickname his generation gave the British Soldier. He wrote it in the quiet, comfortable lull before the Boer War, when politicians and public felt safe, and respect for the Army had declined. But it could have been written two decades earlier when the first Guardsmen arrived at the new Caterham Depot. The relaxed, peacetime mood was intact, and the welcome was hardly ecstatic.

On October 7th 1877, the Croydon Advertiser & Surrey County Reporter announced:

ARRIVAL OF THE REDCOATS
Good News for Servantdom

On Tuesday, the first portion of the 800 troops to be quartered at the new barracks at Caterham arrived at Caterham Barracks to the number of about 300, and thence marched to their new home. The owners of property in the neighbourhood seem by no means pleased with the advent of the military, considering that their property is likely to be depreciated in consequence, but the domestic servants are almost beside themselves with joy.

The report is short. But it speaks volumes about public attitudes of the time.

Within six months, however, the attitude pendulum had swung the other way. There was war in Europe again between Prussia and Turkey and the first tremors of concern were felt in Britain. In March 1878, the Croydon Advertiser reported:

THE NEW BARRACKS AT CATERHAM

At a moment such as the present when the war cloud hangs ominously over our heads, and though sometimes lifting, frequently threatens to burst upon us with wildest fury, it is natural that those services upon whose devotion the maintenance of our honour or interests may depend, should command a greater degree of attention than they receive in more peaceful times.

The Croydon Advertiser's 'greater degree of attention' soon resulted in a fact-finding tour of the Depot, reported in its columns in the purple prose of the time:

Our work of inspection begins. The recruits are at drill, and very well go through their duties. They look remarkably healthy and a sturdy body of young men, and this part at any rate of Her Majesty's warriors-in-embryo, do not merit the sweeping denunciations of our reformer, Mr Holmes. Boys they are not: young they are, and rightly so, but apparently good hardy fellows. The barracks are well arranged, and finely situate in every respect. They are built to accommodate about 300 men, and about that number are established here. The men consist of recruits for the Coldstream Guards, the Grenadier Guards and the Scots Guards, with a sprinkling of old soldiers and non-commissioned officers from various Battalions.

... On the left of the canteen bar is the old soldiers' room, and well the latter seem to be enjoying themselves, some smoking their pipes in calm serenity, possibly revolving in their minds the question of how long it will be before they are called to lead a more active life, others joining in conversation with comrades, whilst the unrestrained and merry laughter told of a light heart within ...

... Generally speaking, the premises are admirably adapted to the purpose for which they are intended. Our visit convinces us that, on the score of quarters, the men who are being trained at Caterham Hill are remarkably fortunate, both as to healthfulness of situation and good and substantial food, a necessary amount of occupation for their time, a fair proportion of leisure and good treatment generally. There are, of course, other sides of the question, but all things considered, the insight into things military which we had on Tuesday drew us to conclude that at Caterham anyway, military life, although not 'all beer and skittles', is by many degrees removed from absolute monotony ...

So the pendulum of public attitude swung. The Guards settled in, and the fears about the presence of the 'brutal and licentious soldiery' impacting disastrously on this rural community, began to dissipate.

But something else happened.

The tradespeople of the small community found that the arrival of the Guards' Depot was accelerating the creation and development of Caterham's commercial enterprises, a process begun when the Southern Railway put its railhead there two decades earlier. Suddenly, the Depot's continuous need for everyday items, from food and hardware to fuel and services of all kinds, was stimulating Caterham's trade as never before.

Within two years, the traders of Caterham were hosts at a dinner for the Depot's sergeants in the Asylum Hotel (to-day the more politically-correct Caterham Arms). And the friendly exchanges continued, as the Croydon Advertiser reported in February, 1879.

Caterham still had its poor and jobless, said the paper, so the enlisted men in the Depot had raised funds through weekly subscriptions and concerts to provide food for needy families throughout that severe winter.

The Guardsmen didn't have to wait for heaven before they got their reward. It came to them.

The Advertiser's next issue reported that their generosity had struck a chord with the local gentry. The well-off locals had got together to pay the annual salary of a Scripture Reader for the barracks! The Guardsmen's reaction is not recorded.

So the friendly traffic went on and the town continued to thrive as it did under the huge expansion of the Depot to meet the urgent recruitment needs of, first the Boer War, and then the two World Wars. Some examples of this are to be found in this book, among the reminiscences of Barney Creton, who was born and raised in the Depot.

Walter D. Gaskin, a local historian, has commented on the social climate of the time in his publication, ***Caterham and the Guards***:

So friendly and tolerant were relations between the civil and military authorities that meetings sponsored by the London Evangelisation Society were allowed in the Depot schoolroom, and in the Caterham Free Press of the 12th February, 1895, we read that the Depot football team played the Caterham Star club on the Barracks ground. During the summer months, cricket and rifle shooting matches were arranged. The officers hunted the countryside around the Depot with the Old Surrey Foxhounds, meeting at the Depot annually until 1912.

The Tillingdown Range was in use at the turn of the century for musketry training. Shooting competitions were arranged with rifle clubs from the South of England, the visitors being entertained afterwards at the Railway Tavern in Caterham Valley.

Miss Florence Lloyd of 'Aldercombe', Caterham, was a leading spirit from 1884 onwards in the provision of a non-sectarian soldiers' home, where recruits might relax in a 'temperance' atmosphere when not on duty. Her struggle was a long one, but eventually two cottages in Caterham High Street, opposite the later soldiers' home and long since demolished, was made available in 1895.

On the 30th April, 1898, the foundation stone of a new building to house the Caterham Soldiers' Home was laid by the Chaplain-General of HM Forces, Rev. Dr. Edghill QHC. The land and building were provided by Dr John Newberry of Meas Mawr, Whyteleafe, at a cost to him of One Thousand, Five Hundred Pounds. These premises have since become 4 High Street and converted into three shops. In his address, Major Frederick, Commandant of the Depot, expressed his gratitude to Miss Lloyd for the number of recruits she had induced to join the Royal Army Temperance Association.

To-day, the chronicled words and the events they describe convey the sense of a vanished world which seems even more remote than the intervening years would indicate.

Gaskin also charts the Depot's construction, and then its gradual expansion and renewal:

The site of the Guards' Depot Barracks, Caterham, was purchased by the War Department on 3rd August, 1875, from John Christopher Bates for the sum of £4,600, with a small plot from a Henry Alston Palmer for £1,712 10s. These together comprised Woodville House, with outbuildings, gardens and grounds covering an area of around 22 acres.
The original barracks were enclosed by walls an all four sides. It consisted of four two-storey blocks, named York, Wellington, Cambridge and Albemarle, a cookhouse, hospital, guardroom, officers' quarters and mess, 'A' Block, married quarters, three drill sheds,

shoemakers' and carpenters' block with a school above, quartermaster quarters, sergeants' mess and canteen, stores, office, workshop, coal yards, latrines and outbuildings. Building was completed by Messrs Higgs and Hill on 10th November, 1877, at a cost of £46,273.

In 1882, one more two-storey barrack block was added (Victoria). The front blocks, then called York, Cambridge and Woodford, were raised to three storeys in 1899 and the other three, Wellington, Albemarle and Victoria, were similarly enlarged in 1900. A recreation establishment, later a quartermaster's stores, was built in 1887 and a theatre opened above the recreation rooms in 1889.

The Guards' Chapel, St Michael the Archangel, was erected in 1886. The War Office provided the site and contributed £3,000, but the cost was met mainly by public subscription, General Philip Smith CB, being the most generous donor. The architect of this well-appointed church, seating 650, was William Butterfield. (Who also designed Keble College, Oxford; the chapel and quadrangle of Rugby School; St Alban's, Holborn; All Saints, Margaret Street, London and, in the antipodes, Perth and Melbourne cathedrals. He was awarded the RIBA Gold Medal in 1884).

Coloured mosaics in the walls of the chapel represent the badges of the first four regiments of the Brigade, the badge of the Welsh Guards not being represented as this regiment had not been formed at the time the work was carried out.

The statue of the Duke of Wellington which stood in the porch of the chapel, had once stood at the end of the barracks parade ground. It commemorated the Duke's close association with the Guards and their decisive role at the Battle of Waterloo.

Between 1889 and the turn of the century, various buildings were added, including the Commandant's house and further married quarters. A skittle alley, built in 1877, was converted to an armourers' shop. Following the Guards' engagement in the Boer War, which culminated in the formation of the Irish Guards, two more plots of land of about 22 acres were added. 'D' Block married quarters was built on one, next to the Tally Ho! public house; the other allowed space for a gymnasium in 1901, and 'C' Block of 14 married quarters in 1904. In the same year, a thirty-yard firing range was built on the gardens of soldiers living outside the Depot. (A second range was to be added in 1925.) The access to the ranges became Magazine Road. Two more barrack blocks, Codrington and Roberts, were added in 1905.

Gaskin records that much of the hurried emergency build-up during WW1, including the huge hutted accommodation complex, rapidly disappeared with the return of peace in 1918. In 1925, the Depot boundary was extended northwards as far as The Grove by the purchase of the 'Tin Town' WW1 encampment. This addition of eight acres raised the Depot area to the maximum size it was to reach until WW2.

The 'twenties saw a steady programme of improvements. Most significant was the introduction of electricity for lighting throughout the Depot in 1928. Gas was retained in the cookhouse and in the hospital, where it was used to sterilise instruments and equipment. The first permanent dining room was built in the same year as part of an improvement scheme covering the cookhouse, boiler house, and bread and meat stores.

This followed a major diphtheria outbreak in the district, which shut the Depot down in two stages. In the first, part of the barracks population was moved to Warley in 1927. A year later, the whole Depot was moved to Canterbury. During the evacuation, the barrack buildings were given a rigorous health check. The drainage system was thoroughly examined. It was found to be in good order.

The Guards returned to Caterham in September, 1929. On medical advice, accommodation of men and NCOs was revised to eliminate overcrowding in the barrack blocks. By the early 'thirties, a garage had been added for an ambulance and other transport, and three blocks of married quarters improved or reconstructed to meet higher living standards. A museum was opened in one of the cricket field huts and, in 1933, a swimming pool was created in the cinema pavilion. The pool could be covered when necessary to allow entertainment, including boxing tournaments.

By 1938, the darkening situation on the Continent put the Depot on a war footing. The old magazine was converted to a chamber for chemical warfare training. Early the following year, the huge increase in the Depot population which would result from the outbreak of hostilities in September 1939, was anticipated by further rapid growth. Another six acres at the end of The Grove were bought by the War Department. A security fence enclosed the site, but no doubt lingering hopes for peace caused it to be levelled and laid out as a cricket pitch and two football pitches.

Gaskin records that the impact of WW2 saw the rapid redevelopment of all such areas for training purposes. Three groups of huts for in-coming militiamen were erected at the east end of the sports ground — 17 barrack rooms and ablutions for 340 men. Huts were hurriedly built on requisitioned land outside the Depot and spare land inside it, to accommodate a training camp for six companies. Two NAAFI canteens, each serving a thousand men, were built, and three combined cookhouses and dining rooms, each for 600 men. In the same period, three gymnasiums, a gas defence centre, two first-aid and medical inspection rooms and a dental centre were built. With a further two thousand men expected, two additional miniature and thirty-yard ranges and two parade grounds were added.

All this urgent work was completed by 1941, despite frequent attacks by the Luftwaffe.

The war required many makeshift, emergency measures to accommodate the training and living quarters for thousands of recruits. These facilities came into existence with crisis-driven speed. But by 1947, with peace restored, they were gradually disappearing. The huts went from the cricket field and the clock tower was removed, the clock being bought by the local authority and installed in the clock tower in Queen's Park, Caterham.

Modernisation continued in the 'fifties. Victoria Block was rebuilt and central heating arrived. A Royal Engineers unit moved out of the militia huts at the sports ground, and the training camp at Coulsdon was demolished. St Michael's Chapel, just inside the main gate, was re-roofed.

By 1956, the last vestiges of the Depot's wartime footing were being removed. All the air raid shelters were demolished. The slit trenches which had sheltered personnel caught in the open during bombing and aerial strafing attacks were being filled in.

In 1960, the Depot moved across Surrey to Pirbright.

The various barrack blocks in the Caterham Depot had distinguished names, whose bearers had significant links with the five regiments of Foot Guards:

ALBEMARLE	—George Monck, First Duke of Albemarle
ALEXANDER	—Field Marshal Harold Rupert Leofric George, Earl Alexander of Tunis
CAMBRIDGE	—Adolphus Frederick, Duke of Cambridge
CODRINGTON	—General Sir William John Codrington
EDWARD	—HRH Edward, Prince of Wales, later King Edward VIII, and then Duke of Windsor
ELIZABETH	—HRH Princess Elizabeth, now HM The Queen
MONCK	—George Monck, First Duke of Albemarle
ROBERTS	—Field Marshal Frederick Sleigh Roberts VC, First Earl of Kandahar, Pretoria and Waterford
VICTORIA	—HM Queen Victoria
WELLINGTON	—Field Marshal Arthur Wellesley, First Duke of Wellington and Ciudad Rodrigo
WOODFORD	—Field Marshal Alexander George Woodford
YORK	—Frederick August, Duke of York and Albany

General View of Barracks with Third Floor Extensions

A Photographic Gallery of the Old Depot

A Scots Guards fatigue party from around 1910 — and the kitchens, ablutions, latrines and stores about to get a good seeing to . . . . . .

In 1910, the Coldstream recruits won the Depot competition for the best mess room Christmas decorations . . .

The Depot's Drum and Fifes led many a ceremonial parade. This was the unit in 1908.

Walking out of an evening, the cane, or swagger stick, was an important accessory before WW2.

In the early 1900s, and throughout the Depot's life, changing guard at the main gate attracted interested spectators

Call to the Colours

Recruiting Day in Caterham . . . . The Great War to End All War had been raging on the Continent since August 1914. In Britain, volunteers had rushed to the recruiting offices in their tens of thousands, but the need for more front line troops was still urgent. These scenes were recorded in Caterham in April 1915 when a Guards detachment marched down Church Hill to a saluting base outside the Railway Hotel. With patriotic fervour, the townspeople turned out to cheer.

This photograph and the one on the following page were taken in August 1914. They show the first impact of WW1 on the Depot. Almost all the recruits are still in civilian clothes and they carry wooden staves instead of rifles. Within a few months, several thousand were under training at Caterham, all of them volunteers - conscription was not introduced until much later.

Another WW1 photograph shows urgent training amid new buildings and tented accommodation - complete with washing lines.

The impact of WW1 and the influx of volunteers meant Depot accommodation had to be expanded rapidly. This was 'Tin Town' under construction in 1914 . . . .

. . . . and ready for recruits.

Part of the Irish Guards' lines in 1914

Women volunteers played an important role at the Depot in both world wars, releasing many men for active service. This group of auxiliaries, photographed at Caterham during the 1914-18 war, were typical of the thousands who cooked, nursed, drove ambulances and other vehicles, and performed clerical duties in the UK and overseas.

Medical Officers and Nursing Staff — 1917

The Hospital

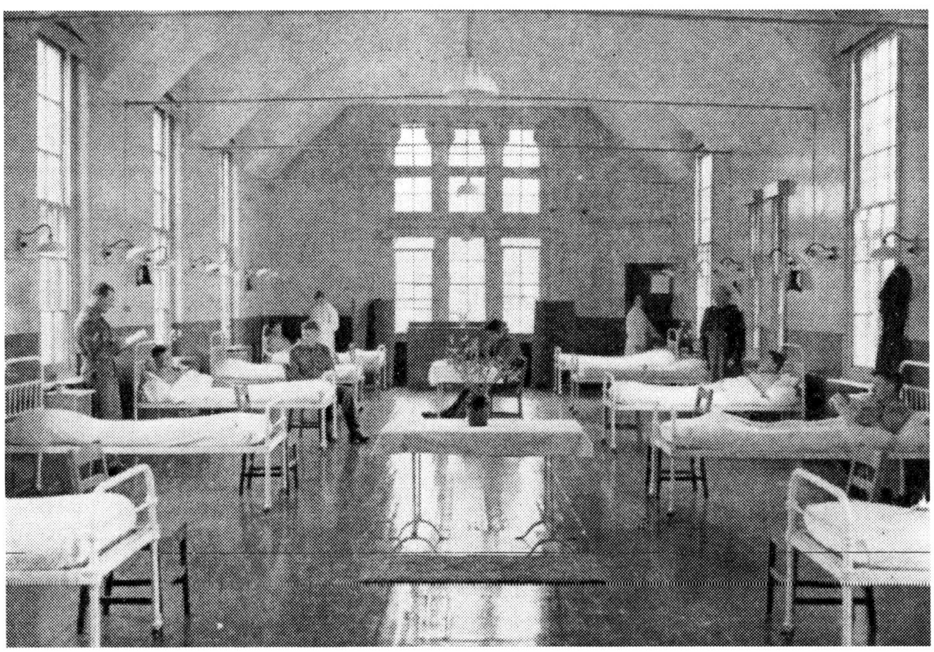

A Scots Guards squad in 1917. After a few weeks in the training battalion they were bound for France and the trench warfare of WW1.

The Depot Orchestra just before WW1

In the 1920s . . .

Recruit training on the miniature range

The gymnasium

. . . . and the hospital

The swimming baths — they could be converted into a theatre

A Grenadier squad at dinner. The oil lamps were not replaced by electricity till 1928..

Commandant's Parade was usually held on Saturday mornings and all regiments took part. This one was on the main drill square in the '30s. With several thousand recruits in the Depot during WW2, the parade was held on the much larger Joliffe Square.

Two aerial photographs of the Depot in the 1930s

Barrack room ready for inspection in the 1920s.

Coldstream Lines.

The Officers' Mess

The Chapel of St. Michael and all Angels.

The social life of the Depot was always strong. It reflected the family and community spirit which linked town and the men of the five regiments. The Corporals' Dance in the flag-decked gym (above) was a popular and regular event, attended by many local girls.

***Christmas Dinner* was one of the family highlights of the year.**

Guards' Depot
Parade Ground

Images from the 1800s

Back in the 1800s, photography was still very much a minority activity. Professionals seized on the commercial opportunities, recording events of the day with their large brass and mahogany tripod cameras, producing glass plate monochrome negatives - hence the early pictures of Caterham Depot life in these pages.

But although photography was not yet available as a popular hobby in the 1800s, it didn't stop men who served in the Depot recording the characters who shared their hard life. Over the years, many produced their own sketches, showing how they saw themselves and their fellow Guardsmen.

Unfortunately, most of these spare-time drawings have been lost, but some have survived in a scrapbook kept by an unknown collector and preserved in the archives at Wellington Barracks, London.

This history consists of recollections of Guardsmen who passed through Caterham in the 20th Century. Something of the spirit of the men who preceded them is in the selection of sketches in the pages which follow.

'The Gargoyle' - an otherwise un-named Depot character

The Organisation of Depot Life

GRENADIER GUARDS
KIT
As laid out for inspection

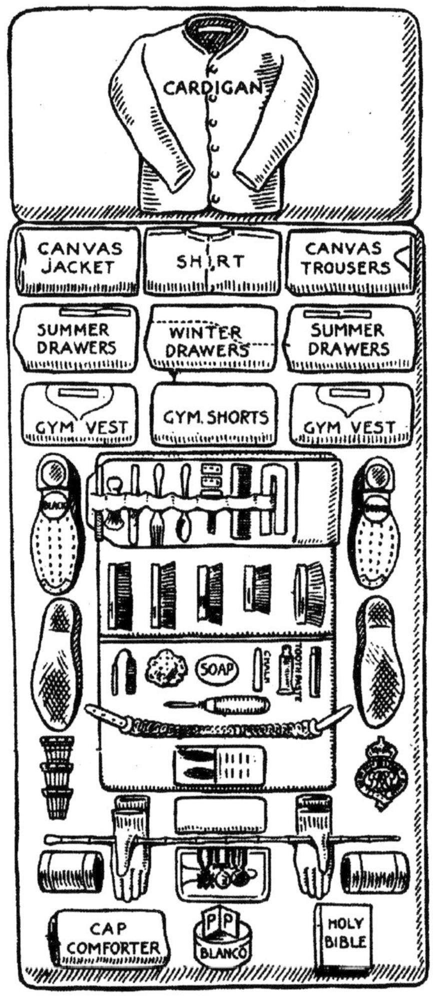

Two pairs of Socks to be hung on left forearm, numbers to the front when arms extended.
In Summer, Cotton Drawers will be in centre, and Winter Drawers on either side.
Two pairs of gloves to be shown.

STANDING ORDERS

OF THE

BRIGADE OF GUARDS

Issued by the Authority of

MAJOR-GENERAL H. TROTTER

COMMANDING THE BRIGADE OF GUARDS

(FOR PRIVATE CIRCULATION)

LONDON
PRINTED BY WILLIAM CLOWES & SONS, LIMITED
STAMFORD STREET AND CHARING CROSS
1899

SECTION XIII.

DEPÔT.

1. ORGANISATION.

701. The Guards' Recruit Depôt is under the command of a Field Officer of the Brigade of Guards, and consists of one depôt company from each regiment, subdivided into two parts for administrative purposes. 789/95

702. These companies are composed of— B.M. 822/97

(1.) The depôt establishment, including—
 (*a*) The staff of the depôt.
 (*b*.) The depôt of each regiment.

These numbers are published in Army Orders.

(2.) The recruits.

703. All appointments to, and changes in, the staff of the depôt are made by Brigade Order. B.M. 822/97

All appointments to, and promotion and changes in, the depôt of a regiment are made by Regimental Order.

704. The 70 trained soldiers on the depôt establishment of each regiment are allotted as follows :— B.M. 822/97

(1.) As assistant drill instructors (lance-corporals)	8
(2.) As trained soldiers	37
(3.) Specially employed	25

B.M. 640/98

Those specially employed under (3) are divided into three classes, as follows :—

G 2

1. Organisation—continued.

(*a.*) Those detailed by Brigade roster of regiments for casual duties at the depôt or by Officers Commanding regiments, or by both.

	No. Detailed by Roster.	No. per Regiment.	Total.
Assistant Instructors in Gymnasium .	..	2	6
School assistants	2	1	5
Clerks	..	1	3
Tailors	1	1	4
Shoemakers	1	1	4
Garrison military police . . .	..	1	3
Pioneers	..	2	6
Butcher	1	..	1
Cooks	1	2	7

(*b.*) Those selected by the Commandant from the trained soldiers at the depôt.

Orderlies	8
Waiters	8
Barrack police	1
Postman	1
Librarian	1
Storeman	1
Wash-house.	1
Gardeners	2
Officers' mess	2
Barrack labourer	1

(*c.*) Those who accompany their masters to the depôt.

Officers' servants	10

705. During the Commandant's absence on leave, his duties will be performed by the next senior officer at the Depôt, who, if not under the rank of Captain, will exercise the full powers of a Commanding Officer.

706. For Brigade questions the Commandant is responsible to, and will receive orders from, the Major-General; for Regimental questions he is responsible to,

1. Organisation—continued.

and will receive orders from, the Officers Commanding the regiments concerned.

707. The Commandant will sign his name as commanding the Guards Depôt.

708. The Commandant is appointed for two years. 464/93

709. The Officer Commanding each regiment will be given, in turn, the opportunity of submitting, for the appointment of Adjutant, the name of an officer for the approval of the Major-General, a roster of regiments being kept for the purpose. Should an officer whose name is submitted not be approved of by the Major-General, the Officer Commanding the next regiment on the roster will be asked to submit a name for approval. 709/93

The Adjutant is appointed for two years. 464/93

710. Company officers are posted to the Depôt for a year.

711. A Subaltern before being posted to the Depôt must be qualified by length of service and experience to take charge of a company. 509/93

712. The Officer Commanding each regiment will be given in turn the opportunity of submitting, for the appointment of Quartermaster, the name of a Quartermaster for the approval of the Major-General, a roster of regiments being kept for the purpose. Should a Quartermaster whose name is submitted not be approved of by the Major-General, the Officer commanding the next regiment on the roster will be asked to submit a name for approval. The Quartermaster will perform all the duties of Acting Paymaster. 1247/93

713. The Sergeant-Major, Quartermaster-Sergeants, Orderly-room Sergeants, Pioneer Sergeant, Sergeant-Cook, and the Canteen and Coffee-bar Managers, will be appointed by rosters of regiments. B.M. 184/97

714. In the case of the Sergeant-Major the Officer Commanding the regiment concerned must undertake that the first Battalion Sergeant-Majorship that becomes vacant after the date of the appointment, will be given to the Sergeant-Major of the Depôt; and the Sergeant-Major must undertake to be removed from the Depôt and to accept the appointment of Battalion Sergeant-Major when it is offered to him. 3/81

Clockwise from top left — progress through the Guards Depot at Caterham. The finished article in the centre.

The Formula

It is an enigma wrapped in seeming contradictions, which has engaged leaders of other regiments around the world.

The Guards' Depot receives raw, sometimes rebellious recruits, accustomed to the freewheeling licence of civilian life. Then it deprives them of much of their liberty, subjects them to rigorous training backed by hard, round-the-clock discipline, and turns out soldiers with fast reactions to given orders as members of combat units, large or small — yet still capable of individual initiative of a high order when the unexpected demands it.

Much of the answer is to be found in the contributions to this book, from ex-Guardsmen of all ranks. In these pages, something the lay reader might well have expected to find is, in fact, totally missing — nowhere is there any resentment against what is generally acknowledged to have been a difficult indoctrination. Instead, there is pride, and a sense of comradeship, of belonging to a unique brotherhood of regiments with a tradition of military achievement spanning four centuries.

There are a number of reasons. One of the most powerful for the individual is the sense of achievement, of having survived and won acceptance in a regime which is not for the faint-hearted or fragile. But it is not all male macho posturing. There is also an element which accepts the values of a system with emphasis on personal accountability within an overall organisation. It is that element of acceptance, that winning of approval for the end result, which distinguishes the training of the young Guardsman from the processes employed in the pursuit of excellence elsewhere in the world.

The following is from the Standing Orders of the Guards Depot, Caterham, dated 1930, and lays down very firm guidelines for drill instructors in all five regiments. These words, from seven decades ago, may be surprising to some. But they define one of the essential factors of basic Guards' training which others find so elusive.

Standing Orders for Drill Instructors

"The squad instructors will see to the drill, cleanliness and discipline of their squads. They will instruct the recruits in all their duties, impressing upon them the importance of discipline and soldierlike behaviour, both in and out of barracks. They should take the greatest interest in their squads, visit the barrack rooms at all hours, and be ever ready to instruct the recruits in all their duties, to give them advice and, by their example, to instil into their minds that high standard of duty for which the Brigade of Guards has ever been distinguished.

*

The squad instructors should, while exacting the strictest discipline, make every allowance for young recruits. They should be careful never to lose their temper, bearing in mind that a man should very rarely be touched in any way, except in the early part of his training, when it is sometimes necessary for the purpose of instruction. It should always be remembered that, with young recruits, many faults are committed in ignorance, and these can be more often rectified by a few words of explanation and advice, than by abuse or punishment. Squad instructors cannot be too particular as to their manner with recruits. A rough manner with men at any time of their service is always a great mistake, but it is far more so with recruits who are just beginning a life which is entirely new to them, and with whom discipline is at first very irksome. It is everything to make a recruit content and satisfied with his life as a soldier during the early part of his training; he is then far more likely to become thoroughly well trained and well disciplined, than if he starts discontented and dissatisfied."

. . . . and among the duties of the Sergeant-in-Waiting ...

"He will see that the windows in each barrack room are opened at 7.45am in winter and at Reveille in summer. He will see that the lights and fires are put out at 10pm. Sufficient windows in each barrack room are to be kept open during the night to provide proper ventilation."

The Transformation Begins

Young Guardsmen absorb the beginnings of the *esprit de corps,* the pride and sense of belonging to their Regiment and the Guards Division, in the early days of their service. It happens gradually, almost imperceptibly. In some cases it happens despite resistance and determination to stay aloof from the ethos of, and sense of belonging to a military organisation with very special qualities and traditions. Reluctant recruits called up during periods of conscription can testify to that.

All the elements of his training contribute to the transformation. From the start, the recruit is influenced most directly and potently by his Squad Instructor and Trained Soldier - the Guardsman who has been selected to look after the squad in their hut-training hours and to back up and encourage his charges by example. Under this unique system, Squad Instructor and Trained Soldier take on personal responsibility for each recruit in their allocated squad, his welfare and his training progress. Good progress reflects credit on both Instructor and TS. The reverse means demerits, and even punishment - 'Lance-Sergeant Smith, you have lost your name. You have an idle recruit!'

But the recruit is also heavily influenced by his observation of the more advanced recruits at the Depot and the skills and standards they have achieved in just a few more weeks than his own fledgling service.

As he progresses, more and more specialists - the armed and unarmed combat experts, the PT instructors, the educational staff, the fieldcraft specialists, and so on - bring their expertise to bear. And always, there is the challenge of the other squads, the other regiments. It is a friendly competition, but no less fierce for that.

The training is concentrated and focused, and when he is not on the drill square, the firing range, or in the gym, the recruit attends lectures in which his officers reveal the basic tactics of warfare, firstly in platoon and company formations. And so he undergoes the slow and sometimes painful conversion to the rank and status of Guardsman, and he leaves the Depot.

Still based on time-honoured principles, to-day's training regime incorporates the lessons of modern warfare - from fullscale, open hostilities to the terrorist's furtive Armalite and bomb attack from hiding. The current methodology reflects the heavy emphasis on operational efficiency, although the traditions and achievements of the Guards regiments are a theme running through the whole programme.

The Depot training used to be much more formalised. In 1895, Major (later Major General) The Hon. HH Henniker wrote Notes for Lectures to Recruits of the Brigade of Guards. The notes date from 1914. To-day their detail and tone are definitely from another age.

These are a few extracts:

FIRST LECTURE...

First Duties, &c.

The Brigade of Guards comprises five regiments —the Grenadier Guards, the Coldstream Guards, the Scots Guards, the Irish Guards, and the Welsh Guards. The Brigade of Guards.

H.M. The King is the *Colonel-in-Chief* of each of the regiments of Guards. The King is of course the head of the Army, but it is only upon a few specially distinguished regiments that he confers the honour of becoming their Colonel-in-Chief. He signifies his special connection with, and interest in, those regiments, by wearing their uniform on certain occasions, and he frequently wears the uniform of his Guards. H.M. The King, Colonel-in-Chief. Organization

Under the King, there is at the head of each regiment a distinguished Field-Marshal or General, who is known as the *Colonel* of the Regiment. The duty of this officer is to watch over the interests of his regiment generally, and, in the case of the Guards, he is usually either a member of the Royal Family, or an officer of exceptional distinction.

Each of the five regiments is commanded (under the Colonel) by an officer called the Lieutenant-Colonel, and is divided into battalions, each of which is commanded by an officer of the rank of Lieutenant-Colonel.

The Grenadier and Coldstream Guards have each three battalions ; the Scots Guards have two battalions, and the Irish Guards and Welsh Guards each consist of one battalion.

Till 1897 the Brigade consisted of three battalions of Grenadiers and two each of Coldstream and Scots Guards.

The Officer Commanding the regiment, therefore, or Regimental Lieutenant-Colonel, commands the whole regiment, and under him are the Officers Commanding battalions of the regiment. A battalion is divided into four companies and the Headquarter Wing. Each company is commanded by a Major or Captain, under whom are a Captain and 4 Subalterns.

The first thing to be realized and remembered by every recruit on joining one of the regiments of the Brigade of Guards is that he should be proud of the honour of belonging to that regiment. No body of troops in the world, now, or in the past, has ever had finer traditions, a more glorious record, or a higher standard than the Brigade of Guards. The Brigade has always set an example of efficiency, discipline, smartness and soldierly qualities both in war and peace which soldiers of our own or any other army should be proud to follow. No man can think too much of this, whether he be an officer, non-commissioned officer or guardsman; for it rests with those at present serving to carry on the tradition and to maintain the standard set up by guardsmen in the past. Any man who falls below that standard brings discredit not only on himself, but on the famous regiment to which he belongs and on the Brigade of Guards of which he is a member. And this high standard which prevails throughout the Brigade exists because every guardsman, of whatever rank, from the highest to the lowest, takes a pride in his regiment, a pride in himself and a pride in his work; and, whatever he has to do, always does his best.

Obedience.

Obedience is the first duty of a soldier. An order given by an officer or non-commissioned officer must be obeyed without hesitation or remark; and soldiers must carry out all orders instantly and cheerfully, however unpleasant the execution of them may be. If a man has anything to say about an order which he has received, if he thinks himself wronged by it in any way, or if he has any complaint to make about it, he must obey it instantly none the less. After the order has been carried out he may make his complaint, if necessary, by asking to see his company officer about it, but in every case he must obey first. It must be remembered that a man who does what he is told, and does his best, always has good friends in the officers and non-commissioned officers over him ; and if he has any good grounds for complaint they will take care to see that he gets redress.

Soldiers will pay the most scrupulous attention to cleanliness. No man is good for much who has not proper self-respect, and no man can be self-respecting if he is dirty. Washing is also necessary to health ; nothing is more healthy than cold water and plenty of it. With regard to the feet, which may become sore from marching, there is nothing better for them than cold water. A good wash every morning and another after the day's work is over will prevent any man from disgracing himself by having to fall out on the line of march. Infantry soldiers must pay the greatest attention to this, as their whole utility depends on their marching power. Cleanliness.

Health, which goes with cleanliness, is most important to a soldier : It is the absolute duty of every soldier to do all in his power to keep himself healthy, for a man who is not strong and healthy can neither march nor fight, and a soldier is of no use unless he can do both. A man, too, who goes to hospital is a nuisance to his comrades, who have to do his duty. Men must therefore take great care of their health. For instance, no man need get ill by loitering about in the cold half-dressed, after getting hot or leaving a hot room, nor by running out in slippers during wet weather. Moreover, men must be careful not to drink too much, for excessive drinking will not only get them into trouble, but will also injure their health, and the same applies to excessive cigarette-smoking, which may ruin a man's constitution. Everything possible is done by drill and gymnastics to make soldiers strong, active, and healthy, but men can do a lot for themselves by cultivating regular and temperate habits. Health.

4

When a soldier feels ill, or when he requires medical treatment of any kind, he will give in his name to the Corporal-in-Waiting at réveillé, in order that he may see the Medical Officer. If he should be taken ill at any other time he will report himself to the Serjeant-in-Waiting. He will not treat himself or consult private doctors, as men are very severely punished for concealing disease, that is to say, for not reporting themselves sick to the Medical Officer.

Men must, however, make it a point of honour not to go sick, unless they are really ill.

Smartness. Soldiers should take the greatest pride in their smartness in every way. It must be their constant thought to turn out not only themselves but also every article of their clothing, arms and accoutrements spotlessly clean and smart. Men are punished for not doing so; but it must be remembered that this is not insisted upon without a good reason. On the contrary, there is the best possible reason for strictness in this respect. Every man who is worth his salt has a self-respecting pride in the way he is turned out; and every soldier takes a proper pride in his appearance in the King's uniform. No man is worth much who is dirty or slovenly in himself or his dress, or in the way he goes about; and no soldier is a good soldier who is not smart. In fact, the smartest soldier will nearly always be found to be the best soldier, whether in peace time or on active service.

Salutes. Soldiers will always be on the look-out to pay the proper mark of respect to their superiors. They will stand to attention, officers and non-commissioned officers will salute, whenever the National Anthem is played, and will always salute the Colours. The orders on this subject are put up in all the barrack rooms, and every recruit must take pains to make himself thoroughly acquainted with them.

Routine. The routine of the Guards Depot is given in the Syllabus and should be explained to recruits.

Men will rise at réveillé, shave, wash and make up their beds. The windows will be opened at réveillé in summer and at 7.45 a.m. in winter. Beds will be made up half an hour after réveillé.

Men will always have their shelves clean and their kits properly put up before parade.

They will take pride in the cleanliness of their barrack-rooms, and, besides keeping their own things perfectly clean and in order, they will be most careful not to dirty the room by upsetting blacking, etc., while they are "shining," or grease while they are having their dinners. All their clothing will be neatly folded on their shelves, all their kit clean, their beds properly made up, their boxes clean and tidy, and everything should be perfectly ready in case of having to turn out suddenly. No turning down of beds will be allowed till after dinner.

Soldiers are responsible for their arms, clothing, and necessaries, which must always be kept perfectly clean.

When anything is required in the way of repairs or alterations they will report it at once; they will not repair or re-mark their accoutrements; they are responsible for the care and repair of their clothing and necessaries, and they will examine every article served out to them, and point out any defect. They will not alter or mark any article themselves, and they are forbidden to sell, exchange, make away with, or lend any articles whatever.

No gambling of any kind is allowed in the barrack-rooms.

Men will be in their barrack-rooms five minutes before tattoo, and will turn down their beds; they will stand to their beds when tattoo roll-call is called, and will be in bed before lights out, after which there will be no talking, noise, or smoking.

Men will invariably be on parade at least five minutes before the time for which parade is ordered.

6

Every soldier should realize the importance of education. At the Guards Depot everything possible is done to help men in this respect and to make it easy for them to better their education.

School. Every recruit on joining must attend school until he obtains a 2nd Class Certificate of Education. It must be remembered that a man's pay depends on his being in possession of this certificate; his promotion depends on it; and the class of employment he can hope for after leaving the colours depends to a great extent on his being well educated.

Both in military and in civil life, therefore, this is of the greatest importance if a man wants to get on, and every recruit must resolve to make the most of his time at school and to work hard.

Discipline. Lastly, it must never be forgotten that the strict discipline under which every man in the army is placed is not insisted on without a good reason. Without discipline it would be impossible to have control over any body of men in action, and the strict discipline in regard to small matters in barracks and on parade is the means of teaching the absolute and immediate obedience which is the basis of discipline that gives a commander control of his men in action and that wins battles.

The object of discipline is to develop a soldierly spirit; to help the soldier to bear fatigue, privation, and danger cheerfully; to imbue him with a sense of honour; to give him confidence in his superiors and comrades; to increase his powers of initiative, of self-confidence, and of self-restraint; to train him to obey orders, or to act in the absence of orders, for the advantage of his corps in all conditions; to produce such a high degree of courage, disregard of self, and confidence in his armament, that in the stress of battle he will use his brains and his weapons coolly and to the best advantage; to impress upon him that, so long as he is physically capable of fighting, surrender to the enemy is a disgraceful act; and, finally, to teach him how to act in combination with his comrades

7

in order to defeat the enemy. As soon as the recruit joins he should be brought under influences which will tend to produce and increase such a spirit, and it is the duty of all officers and non-commissioned officers to assist in the attainment of this object by their conversation and example.

In the 1800s

On Buck House Duty

Grenadier Bill White on duty at Buckingham Palace in the 1930s. In those days, the sentry boxes were outside the Palace railings and sentries had to patrol their 'beat' through sightseeing crowds. Growing numbers of onlookers (and some provocative behaviour) caused sentries to be moved into the Palace yard. Bill White was a Trained Soldier at Caterham from 1932 to 1936.

Trained Soldiers' Rules

At Caterham, it took no time at all for recruits to learn that questioning the authority of their Trained Soldier would put them as deeply in the proverbial as would challenging the Commandant. His word was law. But although he had absolute control in the barrack room, the Trained Soldier had to work with the squad instructor to achieve the transformation of recruits into Guardsmen, while working under strictly-imposed rules.

These rules are illustrated by a memo issued in 1930 to Trained Soldier William White, Grenadier Guards, by the Depot Commandant, Major Trew, Coldstream Guards. It was marked Confidential.

Memorandum to All Barrack Room Trained Soldiers

The efficiency and tone of happiness of the Guards' Depot greatly depends on you personally. As far as the recruits are concerned, you are in the position of a Non-Commissioned Officer.

Think of the time when you were a recruit and you will remember that you worked willingly for the Trained Soldier who led you and did not drive you; also that you had respect for and had confidence in the Trained Soldier who showed himself to be your friend as well as your superior officer.

Being a friend does not mean being familiar; on the contrary, it necessitates keeping the strictest discipline. But no good recruit minds discipline provided that it is fair and impartial and does not vary.

Always set the highest standard possible.

I am most anxious that the Guards' Depot should be a happy place and this can only come about by mutual confidence and respect between all ranks. Do not forget to give a word of praise and encouragement when due.

I want you to realise the tremendous opportunities as well as the responsibilities which are yours and which you can use for the goal of the BRIGADE OF GUARDS.

I wish to impress upon you the extreme importance of avoiding the use of obscene language and any suggestion of laying violent hands on recruits.

Any such cases brought to light will be severely dealt with.

In the 1800s

'The Little Alabama Pullet' - a popular turn among the Depot's home-grown entertainments.

The Voice of the Bugle

Daily barracks life is ordered by a repertoire of bugle calls, the origins of which are buried deep in centuries of tradition. Recognition and understanding of these calls are priorities in the steep learning curve recruits undertake when they arrive at the Depot. They open and close the day, summon to parade for drill or punishments, to guard duty, fatigues and meals; they warn of emergencies, like fire and attack. To the recruit, all are urgent and vital to keeping out of trouble — or to not getting in deeper if he's already in it.

To fix calls indelibly in the memory, each has a piece of ancient doggerel which matches the phrasing of the bugle's notes, and conveys a rough sense of the call's message. Many would have brought a blush to a maidenly cheek before the days of TV drama, but the words today, like the calls, were old when the Guards were winning battle honours in the Napoleonic and Crimean wars.

These are a few of the calls in regular use —

POST CALL: Letters from Lousy Lou, boys, letters from Lousy Lou!

OFFICERS: Officers come to headquarters, officers come to HQ; do do do!

PIONEERS: Pioneers! pioneers! pioneers! The dogs have shit on the square!

TAPS or QUARTER CALL: You've got a face like a chicken's arse!

DISMISS or NO PARADE: There's no parade today, there's no parade today!

PARADE FOR GUARD: Come and do a piquet, boys, come and do a guard; be on time or else, boys, it's seven days hard!

FATIGUE PARADE: I called him, I called him, the bastard wouldn't come; I called him, I called him, he's in bed with Chelsea Lil!

DEFAULTERS: You can be a defaulter as long as you like, as long as you answer your name!

SICK CALL: Sixty-four, ninety-four, he'll never go sick no more; the poor bastard is dead!

ALARM: The enemy's about, the enemy's about!

FIRE ALARM: There's a fire, there's a fire; run, you bastards, run, you bastards, run!

OFFICERS' DINNER: Officers' wives get pudding and pies, while sergeants' wives get skilly!

ROUSE: Get out of bed, get out of bed; you lazy bastards, get out of bed!

ORDERLY ROOM or MEMORANDA: Memoranda, memoranda; be sharp for memoranda; you're in a state, you're going to be late, the CO's waiting for you; justice will be done!

FALL-IN: Fall-in A, fall-in B; fall-in all the company!

COOKHOUSE: Come to the cookhouse door, boys; come to the cookhouse door!

Most of these calls, and many others, used to float out every day over Caterham-on-the-Hill.

In the old days, local clocks were set by them.

GUARDS DEPOT CORPS of DRUMS
August 1931

Drummers H. Wilkins, A. Stephenson, A Chandler
Drummers E. Christian, J Edwards, C. Frost, J.Shea, C Hockney, L/Cpl. C. Mills, Dmr. A. Hollier
Dmr. F. Harris, Lt. Treadwell, L/Sgt. W.Tilbury, Dmr. T. Jones

Queen's Regulations (She was not amused)

In the Guards, punishments — like leave passes, credits for especially good turnouts, and decorations — are 'awarded'. The word conveys nothing of the shattering experience of being marched into the company office at high speed, to stand rigidly at attention with eyes fixed on a point above the CO's head, listening to the sergeant-major describe the alleged offence in a voice fit to rattle the bars in the guardroom cells.

Even in to-day's liberal climate, Queen's Regulations — the Army's codified basic guide to military law — is something no soldier wants to investigate on the basis of personal experience. But while such 'awards' still have the sting necessary to bolster and maintain essential discipline, punishments now are nothing like those imposed last century. Then, the adage 'tempering justice with mercy' hardly arose in the military mind.

Until the second half of the 1800s, deserters were liable to be branded when caught. At the beginning of that century, flogging was a regular punishment for serious offences and was carried out before the assembled battalion. Making the necessary arrangements was one of the responsibilities of the regimental sergeant-major, although he didn't carry out the punishment himself. It is said that it was the drum major's responsibility to count the required number of strokes. A surgeon-officer was always present to observe and, if necessary, to curtail the punishment in order to save the unfortunate's life.

In some cases, judged to be particularly heinous, it was even possible that the sentence could require a thousand strokes of the lash. In such cases, the agony was frequently protracted by being administered in instalments of, say, 50 lashes per day in order to avoid the recipient's death. And always, it was witnessed by the assembled battalion in which the offence had taken place.

Lesser crimes could still mean a term of penal servitude (imprisonment with hard labour) — pointless, arduous physical work directed totally at punishment and deterrence which had nothing to do with reforming the wrongdoer. Under military law in the last century, penal servitude as 'field punishment' sometimes meant the convicted man digging a hole, digging a second hole, and using the spoil from that to fill in the first hole. And so on, *ad infinitum,* or until the sentence was completed.

That standard of punishment was common to the whole Army in the last century. In the 1990s, military justice has tended to follow reforms in its civil counterpart. In the Guards, punishments have kept pace with the steady flow of reforms of the century.

Until the 1950s, however, one punishment was peculiar to the five Guards regiments. It was an astonishingly bizarre 'award' from Queen Victoria herself. It is said that she was upset by one indifferently turned-out Guardsman on duty at Buckingham Palace, so much so that she decreed that all five regiments in the Guards Brigade should undergo an early morning punishment parade for one hundred years.

Over the years, it became known as Breakfast Parade, but it meant that each day, an item of equipment had to be paraded and shown as fully up to the high Brigade standards.

One day it could be boots shown, suitably polished; another, all webbing equipment properly blancoed; yet another, blankets paraded for shaking (the military alternative to dry cleaning), and so on and on.

Under Victoria's sentence, the timetable went like this:

Reveille — 0600 hrs
Breakfast Parade — 0630 hrs
Breakfast — 0700 hrs

Very few Guardsmen were excused Breakfast Parade — tailors and cooks, and not many others. So, from the '50s onwards, when the 'sentence' ran out, Guardsmen enjoyed the relaxed luxury of having an hour to perform their ablutions, make up their beds and tidy their bedspace between reveille and breakfast call.

The 'Menu' of Punishments in the 1920s included:

SCALE OF PUNISHMENT

Offence	*Punishment*
Absence not exceeding 15 minutes	1 extra parade.
Over 15 minutes and under 30 minutes	1 drill.
For every half-hour or part of it over 30 minutes	1 additional drill.
From 4 to, and not exceeding, 24 hours	C.B., 2 to 7 days.
Exceeding 24 hrs.	C.O.
Allowing a prisoner to escape ...	C.O.
Altering clothing without leave ...	C.B., 2 to 7 days.
Attempting to go out of barracks when a defaulter or ordered to remain in	C.O.
Bad order for guard or picquet ...	3 to 5 days' C.B. or 2 extra guards or picquets.
Bad order for parade	Drills or C.B. at discretion of Company Commander.
Bed, ill made up	Show up, time to be stated.
Being in a prohibited public house	C.O.
Being in possession of kit the property of a comrade	C.O.
Borrowing without authority ...	C.O.
Breaking away from escort ...	C.O.
Breaking out of or into barracks	C.O.
Coming into barracks in a dirty state	C.B., 3 to 7 days.
Clothing in bed	1 extra parade.
Creating a disturbance after Lights Out	C.O.
Creating a disturbance in a barrack room	C.B., 3 to 7 days.
Dirty bed-space	1 extra fatigue and show clean.
Dirty articles for parade	1 extra parade each article same order.
Dirty articles for Public Duties ...	Double punishment.
Dirty flesh for parade	2 drills.
Disobeying an order	C.O.
Discharging a live round unnecessarily	C.O.
Drunk in barracks	C.O. Fine as relevant.
Drunk on or for parade	C.O. Fine as relevant.
Drunk in the streets	C.O. Fine as relevant.
Fighting in barracks or camp ...	C.O.
Fighting in the streets	C.O.
Firing on the wrong target ...	2 extra fatigues or drills; 2nd offence, 3 days' C.B.
Gambling	C.O.
Going into the Canteen at improper hours or when not allowed in	2 drills; 2nd offence C.O.
Grumbling or arguing unnecessarily with a senior	C.B., at discretion of Company Commander.
Having a bruised face	Remain in till well, except under exceptional circumstances.
Idle at drill	1 drill.
Idle on sentry	2 drills.
Idle swabbing or fatigues ...	1 extra fatigue.
Idleness in carrying out an order	3 drills.
Improper conduct on sentry—deserting his post, sitting down, sleeping or gross negligence	C.O.
Improperly dressed in the streets	C.O.
Inattentive at drill	1 extra parade.
Incompetence or inefficiency—continual offences by Non-Commissioned Officers	Forfeiture of privileges at discretion of Company Commander.
Incompetence or inefficiency by Non-Commissioned Officers (of a gross nature)	Forfeiture of privileges at discretion of Commanding Officer
Insubordination	C.O.
Kit badly set up in the barrack room	Show kit properly up, time to be stated.
Kit in bad order	3 drills and show kit.
Late for parade	1 extra parade and upwards according to circumstances.
Leaving kit about in the barrack room	1 extra fatigue.
Leaving the ranks without leave ...	C.O.
Losing a permanent pass ...	Pay for a new one.
Losing necessaries	Replace them.
Making a frivolous complaint ...	Drills or C.B.—at discretion of Company Commander.
Making away with clothing or necessaries	C.O.
Missing—	
Blanket shaking	1 Drill.
Change of Quarters	C.O.
Commanding Officers' Memoranda	C.O.
Company Officers' Memoranda	2 drills.
Drill when C.B.	C.O.
Falling in with Defaulters ...	2 drills.
Falling in with Picquet ...	2 drills.
Fatigue	2 drills.

Guard, wilfully, though in barracks	C.O.
Parade	C.O.
Paying extra parade	2 drills.
Picquet, wilfully, though in barracks	C.O.
Seeing the Medical Officer ...	2 fatigues and remain in till seen.
Swabbing	2 extra fatigues.
Tattoo, if in barracks	2 extra fatigues.
Not having his bed made up proper time	Show up at time to be stated.
Not obeying the orders of a sentry	C.O.
Not obeying the orders of his post	C.O.
Not knowing the orders of his post when offence does not amount to gross negligence	2 drills.
Not shaved for parade	1 drill.
Obtaining leave under false pretences	C.O.
Offences out of barracks	C.O.
Quitting fatigues or parade ...	C.O.
Quitting swabbing	1 to 4 extra fatigues.
Resisting an escort	C.O.
Rifle—	
Safety catch not applied ...	1 extra parade.
Dirty	2 drills.
In bad order	3 days' C.B.
Slovenly dressed for guard ...	2 extra parades, same order.
Slovenly dressed for parade ...	1 extra parade, same order.
Talking in the ranks	2 drills.
Talking on sentry	According to circumstances.
Traffic accidents	At discretion of Commanding Officer subject to local Brigade or District Standing Orders.
Urinating in an improper place ...	C.O.
Using disgusting or obscene language	According to circumstances.
Using improper language to a Non-Commissioned Officer	C.O.
Wrong order for duty, parade or fatigues	1 to 3 extra parades, same order.

FORFEITURE OF PASSES

Regimental Entry	Plain Clothes and Permanent Passes forfeited for one month.
Company Entry	Plain Clothes and Permanent Passes forfeited for two weeks.
Non-Commissioned Officers ...	At discretion of Commanding Officer.

This scale is a guide to maintaining uniformity in the Regiment. Obviously, punishments may vary in different theatres and under different conditions of service. No man will be brought before the Commanding Officer till the Company Commander has investigated the case in the man's presence.

After hearing the evidence, and before settling his report, an Officer will invariably ask the man what he has to say in his defence.

Dealing with the Demon Drink!

ORDERS FOR BARRACK GUARD.

1. Men on Guard will be allowed to have one pint of beer each for their dinners and suppers, but on no account at other times.

2. No man of the Guard is to be permitted to fetch beer from the Canteen. The Canteen waiter will take beer to the Guard at 12.30 hours and 19.30 hours.

3. Two men of the Guard will always be told off as patrols, who will take escort duty and look after men in close arrest. These patrols will always be present when men in close arrest leave the detention room for any purpose, when they eat their meals, when they are shaving, and when an Officer visits them.

4. If the man confined is wearing a tunic it should immediately be taken from him and a service dress suit sent for.

5. When a man is confined he will be taken straight to the detention room, or to the cells, by the patrols or the escort. There will be no halt of any kind in the Guard room. The space between the Guard room door and the detention room, or the cells, will be kept clear.

6. All men coming into barracks drunk will be confined. Both non-commissioned officers of the Guard must see them. Should a soldier be handed over as drunk who, in the opinion of the Sergeant of the Guard, is not drunk, the latter will at once report to the Drill-Sergeant-in-Waiting.

7. A soldier detained for drunkenness will be placed, if possible, in a separate cell, where his boots and his belt will be taken off by the patrols. his collar will be unhooked and his jacket unbuttoned. If violent he will be handcuffed.

8. If, from want of accommodation, two or more drunken soldiers are locked up together in the detention room, a sentry in side arms will be posted inside.

9. Men confined for fighting must not be locked up together.

10. Soldiers in close arrest will be visited by the Sergeant of the Guard at least every hour, and those who are very drunk oftener.

The Royal Links

In the 1800s . . .

The Surgeon-Major — a G & T in the mess before dinner.

The Royal Links

The relationship between the Royal Family and the Guards is long-standing and much more than the formal designation of the Brigade as the Monarch's 'Household Troops'. One illustration of this is the career of Prince Arthur, favourite son of Queen Victoria.

Born in Buckingham Palace on May Day, 1850, Arthur was the third son of Victoria and Prince Albert. The indications of his later career emerged at an early age. He decided to make the Army his career and entered the Royal Military Academy, Woolwich, in 1866. He was commissioned in the Grenadiers and by 1893 had attained the rank of General. Ten years later he was a Field Marshal. His mother created him the Duke of Connaught and Strathearn in 1874. He served in Egypt in 1882, and commanded the 1st Guards Brigade at Tel el Kebir. Prince Arthur then went to India and held the Bombay command from 1886 to 1890. Back in England he held a number of other military appointments — Commander-in-Chief in Ireland between 1900 and 1904, Inspector General of the Services between 1904 and 1907, and Commander-in-Chief in the

THE BOY . . .
From Victoria's album —five year old Prince Arthur in Guards uniform, complete with bearskin and miniature musket.

. . . AND THE MAN
Grenadier and career soldier — Prince Arthur, now Duke of Connaught and Strathearn, at Caterham in 1899.

Mediterranean from 1907 till 1909.

He was Governor-General of Canada from 1911 till 1916. He withdrew from public life in 1928. In 1879 he had married Princess Louise Marguerite of Prussia.

Their daughter, Margaret, married the Crown Prince of Sweden (later, King Gustav VI) in 1905. Prince Arthur died in 1942.

King George V inspected unsquadded and newly-joined recruits when he visited Caterham in 1926.

Prince Philip at one of the reviews he carried out at Caterham

Inspection of 6th Bn. Grenadier Guards by H.M. The King, Colonel-in-Chief.

27th MAY, 1942.

W.O's, S/SGTS. and SGTS.

Sgt. E. Berry, C. Entwistle, R. Hudson, F. Lawrence, E. Smith, J. Challinor, E. Everitt, C. Dalton, J. Evason, C. Workman, W. Dalkins, J. McNaught

Sgt. H. Warden, C.Q.M.S. P. Darby, Sgts. A. Martin, S. Cox, W. Wilson, W. Heard, I. Charlesworth, W. Lloyd, T. Wedge, S. Buffee, A. Mansbridge, W. Nash, D. G Eggleton, O. Tilson

C.Q.M.S. H. Platt, Sgt. C. Porteus, F. Marlow, W. Amor, H. Bennett, S. Hill, R. Turner, L. Wilson, E. Bolan, J. Dewsbury, C. Harrison, J. Bradbury, T. Grogan

C.S.M. E. Warburton, C.Q.M.S. O. E. Lewis, C.S.M. J. Laird, C.Q.M.S. D. Beadle

C.Q.M.S. J. Evers, C.S.M. J. Atkinson, A. Everett, A/D.Sgt. G. Hackett, R.Q.M.S. C. Hutchinson, R.S.M. W. Cutts, D/Sgt. F. Dowling, C.S.M. W. Wallace, D. Barton, L. Burrell, C.Q.M.S. C. Knaggs,

Capt. A. H. Penn, M.C., Lt.-Col. A. F. L. Clive, M.C., Major W. H. Kingsmill, M.C., Capt. & Adjt. The Master of Forbes

Her Majesty The Queen, Her Royal Highness Princess Elizabeth, His Majesty The King, Colonel J. A. Prescott, Her Royal Highness Princess Margaret-Rose,

In this WW2 photograph, King George VI, the Grenadier's Colonel-in-Chief, was accompanied by Queen Elizabeth, Princess Elizabeth and Princess Margaret on a visit to Caterham to inspect the Regiment's 6th Battalion.

H.R.H. Princess Elizabeth — **Colonel, Grenadier Guards**

WITH THE MEMBERS OF THE STAFF AT THE GUARDS DEPOT. 8th JULY. 1949.

G. & H. BUNCE — CATERHAM

Dmrs. Brooks, Lippiatt, Howard, Hall. Bourne, Gallacher. Td.S. Riley, Dmr. Cox. Td.S. Jones, Green Dmr. Dutch
Td.S Kelly, Vernon, Graves, Fox, Egginton, Gunn, Mead, Haynes, Greenwood, Carr, Gilbert, Smith, Phillips, Burns
Cpls. Price, Deering, Wright, Td.S. Brown, Cooper, Jones, Wheal, Stableford, Young, Morss, Ashley, Crawley, Evans, Howard
Cpls. Drake, Bailey, Williams, Lee, Cowan, L/Sgts. McMahon, Whitaker, Randle, Snape Black, Cpls. Rawn, Hines, Hilsdon, Cooper, Wiles
C.S.M. J. Collins, C.Q.M.S. R. Boyles
L/Sgts. Fisher, Doyle, Hutchings, Sgts. Kingdon, Parry, Gilman, Charlton, Gledhill, Wakelin, Mayhew, D/Maj. Payne, L/Sgts. Biggenden, Argent, Fee,
W.O. F. Watson, R.Q.M.S. J. Williams, Capt. & Q.M. H. Lucas, M.B.E., Lieut H. Hamilton-Dalrymple, D/Sgt. W. Smart, C.S.M.I.M. R. Ankers
Lt.-Col. J. Chandos-Pole, Col. R. H. Goulburn, D.S.O., Maj. R. G. C. Rasch, H.R.H. Princess Elizabeth, C.S.M. F. Giles, Lt. Gen. Sir F. Browning, K.B.E., C.B., D.S.O., Maj. E. G. W. M. Fenn, M.C.

Colonel of the Welsh Guards, Prince Charles inspects the regiment at Caterham on St. David's Day. His escort is the Commanding Officer, Lieutenant Colonel PRG Williams..

Another WW2 visit by ***Princess Elizabeth*** *when she made the day for 114 Guardsmen by sitting in this group photograph.*

GUARDS DEPOT, CATERHAM

At Home

SATURDAY, 15th JULY, 1950

The Open Door Policy

The Open and Family Days

Every Commandant who took over the Depot knew he was responsible for maintaining standards expected of the Guards, set by centuries of tradition. But apart from the demanding rigour of recruit training, he had another task which required no small skills of diplomacy and tact.

To-day, it would be described in the buzz words, 'public relations' or 'good neighbour policy'. Simply, it was just recognising the importance of getting on with the civilian population of the town.

In the main, it was a task well done. Of course, everything was not always sweetness and light in the relationship. At times of national crisis, there were thousands of recruits in the Depot and strains were inevitable. But the history of the Guards and Caterham is a surprisingly peaceful one.

Keeping the peace was an on-going, day-by-day concern for the Commandant. But back in the 30s, someone hit on the idea of an annual event, at which the Depot could be accessible to the ordinary folk of Caterham.

So began the tradition of Open Days, when civilians could visit the Depot, inspect the barracks and its facilities, and watch the recruits demonstrate their skills. It was an instant success.

Within a few years, the annual Open Day had become a popular event, not just for the people of Caterham, but for many others throughout the UK.

Because the Open Days quickly became a social fixture, their character changed to include much more music by the various Guards' regimental bands, and groups of the bands' musicians playing popular light pieces and even jazz. By the 50s, the programme had expanded to include the London District Inter-Unit Athletic Championships.

Some examples of At Home Programmes are shown on the following pages.

The Horseguards Parade
Whitehall

GUARDS DEPOT, CATERHAM

At Home

SATURDAY, 15th JULY, 1950

Guards Depot, Parade Ground

Programme

★

11.15	CRICKET MATCH Guards Depot *v.* Household Brigade
1.30	DEPOT OPEN TO VISITORS
1.45	DRILL DEMONSTRATION
1.45	COMMENCEMENT OF THE LONDON DISTRICT INTER-UNIT ATHLETIC CHAMPIONSHIPS, 1954
3.30	SWIMMING DISPLAY—Swimming Bath
3.45	PRESENTATION OF PRIZES BY MAJOR-GENERAL G. F. JOHNSON, C.B., C.B.E., D.S.O., General Officer Commanding London District and Major-General Commanding Household Brigade
4.0	GUARD MOUNTING—from Irish Guards Square
4.30	TEA INTERVAL
4.30 (approx.)	PHYSICAL TRAINING DISPLAY—on the Cricket Field during the tea interval of the Cricket Match
5.45	MASSED RETREAT—Barrack Square
6.30	SWIMMING DISPLAY—Swimming Bath
7.0	ORGAN RECITAL—Mr. W. H. Milnes, F.R.C.O.—Chapel
7.30	EPILOGUE SERVICE—Chapel
8.0	DANCE (for Adults only). 8 p.m. to 11.45 p.m.—Barrack Gymnasium. Admission : Civilians 2s. 6d. Forces in Uniform 1s. 6d.

(See page 23 for notice regarding late trains and buses.)

GUARDS DEPOT
1950

TRAINING OF THE RECRUIT

On some of the pages of this programme there are photographs of the Guards Depot. Recruits' training lasts sixteen weeks, which is not a long time in which to turn out a smart, well-disciplined soldier: each day and hour of this time are carefully planned so that he may be fit to join a Battalion of his Regiment when he leaves. At the Depot he is mainly taught drill and discipline. More advanced training is done when he reaches his Battalion.

On his arrival at the Depot, the recruit is first taken to the Receiving Room, where he spends a day or two. During this time he is seen by the Commandant, the Padre, the Medical and Dental Officers. After this, when he has been passed physically fit, he goes to his Company, where he waits until a squad of about twenty recruits is formed. A Sergeant or Corporal is the squad instructor, who is helped by an experienced Guardsman known as the Trained Soldier. The latter teaches the recruit how to look after and clean his equipment and is an adviser on many matters. The squad stay together throughout their training.

This training consists largely of drill, which develops the recruit physically, gives him the bearing of a Guardsman and teaches him to take a pride in doing everything really well and as part of a team. The Regimental and squad spirit is strongly encouraged.

P.T. in the Barrack Gymnasium plays an important part in the training programme. There is a swimming pool with warm water for use all the year round, and there are many playing fields. As a result, many recruits put on more than a stone in weight during their training and grow in height as well.

Besides drill and P.T., the recruit learns to handle simple weapons. He has lectures on such subjects as Regimental History and National Savings, besides several talks by the Padre.

Education is not forgotten in the training, and in over forty periods he is taught mathematics and English, besides having lectures on citizenship.

After four weeks' training the squad is inspected by the Adjutant. This is the first obstacle in the recruit's training, and earns him his first week-end leave.

The Adjutant's second inspection comes in the eleventh week. After this the worst obstacles are past. During the last weeks he improves the standard already reached, and is inspected by his Company Commander.

Then comes the climax—the final inspection by the Commandant. The recruit leaves soon after and goes to Pirbright or to a Battalion, where he immediately goes on his first well-earned ten days' leave.

What can the recruit do in his spare time? Besides the playing fields and swimming bath mentioned earlier, there are dances nearly every week, a billiard room, reading rooms and a library. There is a museum where he can learn about the past history of his regiment. There are canteens and restaurants, to which he may take visiting friends and relatives for light refreshments. The Recruits' Reading Room has recently been fitted with a television set.

There is a most up-to-date Dental Centre and a Hospital (M.R.S.), providing every facility for nursing the sick. An experienced Nursing Sister is in charge under the direction of the resident R.A.M.C. Medical Officer.

Finally, there is the Depot Chapel, where the recruit can go at any time to worship and meditate in peace.

This short account will give you an idea of the training the recruit undergoes and the facilities available to him on and off duty. Here he has learnt the essentials of discipline and elementary training. When he leaves as a Guardsman at last, he has the knowledge which will stand him in good stead later in his service. It has been hard work, and at times he may not understand the need for all he has to do, but he is rightly a proud man when he finally leaves the Depot.

This training at the Depot instils into Guardsmen the fine discipline which has largely been responsible for the great name the Brigade of Guards has earned for themselves both in peace and war.

LONDON DISTRICT INTER-UNIT
ATHLETIC CHAMPIONSHIPS
1954

Details of Events

Event	*Team*	*Method of Scoring* (Based on five teams)	*Remarks*
4×110 yds. ...	4	20, 16, 12, 8, 4	Run on the "exchange" baton system
4×220 yds. ...	4		
4×440 yds. ...	4		
4×880 yds. ...	4		
1 Mile team ...	4	20, 16, 12, 8, 4	Run as a team—all to count
3 Miles team ...	4		
High Jump ...	2	10, 8, 6, 4, 2	Distance/height of both strings added together. Hurdles are run as "shuttle" relay
Long Jump ...	2		
Weight Put ...	2		
Hammer Throw	2		
Discus	2		
Javelin	2		
Pole Vault ...	2		
120 yds. Hurdles	4	20, 16, 12, 8, 4	
Tug-of-War (Light-weight—88 stone)			

1953 ATHLETIC CHAMPIONSHIPS
London District

Winners: Guards Depot
Runners-up: Guards Training Battalion

GUARDS DEPOT

OFFICERS, WARRANT OFFICERS, and C.Q.M. SGTS. SERVING AT THE GUARDS DEPOT, 1954

Commandant	Lieut.-Colonel I. W. Gore-Langton, M.B.E., Coldstream Guards
Second-in-Command	Major A. G. Way, M.C., Grenadier Guards
Adjutant	Captain P. J. C. Ratcliffe, Grenadier Guards
Assistant Adjutant	Captain T. C. Richardson, Grenadier Guards
Officer Commanding Brigade Squad ...	Captain I. M. Erskine, Grenadier Guards
Quartermaster	Lieut. S. R. Watts, Scots Guards
Regimental Sergeant-Major...	R.S.M. D. H. Whyte, Scots Guards
R.Q.M.S. i/c Clothing	R.Q.M.S. A. Curling, Coldstream Guards
R.Q.M.S. i/c Barracks	R.Q.M.S F. Farnhill, D.C.M., Coldstream Guards
O.R.Q.M.S.	O.R.Q.M.S. J. Euling, Grenadier Guards
Drill Sergeants	D./Sergt. A. Dobson, Grenadier Guards
	D./Sergt. J. West, Scots Guards
W.O. i/c Musketry	C.S.M.I. A. Groom, Coldstream Guards
N.C.O. i/c Messing	C.Q.M.S. G. Woodland, Coldstream Guards
W.O. i/c P.T.	S.M.I. G. Riley, A.P.T.C.

14th COMPANY, GRENADIER GUARDS

Major R. H. Heywood-Lonsdale, M.B.E., M.C.

Captain R. F. Birch-Reynardson	2/Lieut. W. E. Rawson-Shaw
C.S.M. M. Penney	C.Q.M.S. F. Dowland, B.E.M.

1[illegible]th COMPANY COLDSTREAM GUARDS

Major A. B. Pemberton

Lieut. F. W. R. Fisher	Lieut. D. F. Myddelton
C.S.M. V. Dodds	C.Q.M.S. T. Witton

"K" COMPANY SCOTS GUARDS

Major P. J. H. Leng, M.C.

Lieut. C. St.J. G. Moncrieff	2/Lieut. N. Mrosovsky
C.S.M. J. Forsyth	C.Q.M.S. C. Douglas

No. 5 COMPANY IRISH GUARDS

Major P. Doyle

Lieut. A. J. D. O'H. Moore	2/Lieut. W. J. Hall
C.S.M. J. Grant, M.M.	C.Q.M.S. M. Teahan

No. 8 COMPANY WELSH GUARDS

Major W. H. R. Llewellyn, M.C.

Lieut. A. M. W. Severne	2/Lieut. D. G. P. M. Morris
C.S.M. G. Wilkinson	C.Q.M.S. S. Moseley

CHAPLAINS

The Rev. D. S. Coey, M.A.	Senior Chaplain, Church of England
The Rev. B. Tattersall	Methodist Chaplain
The Rev. M. Collins	Roman Catholic Chaplain
The Rev. B. Pratt	Church of Scotland Chaplain

MEDICAL STAFF

Lieut. G. C. Timbury	R.A.M.C.
Lieut. R. Davison	R.A.M.C.
Major L. M. Flower	Q.A.R.A.N.C.

PERSONNEL SELECTION OFFICERS

Captain J. J. Plain	R.A.C.
Captain R. L. Wilkinson	R.I.F.

DENTAL OFFICER

Lieut. E. S. Scott	R.A.D.C.

EDUCATION STAFF

Captain E. Franklin	R.A.E.C.
W.O.II A. Bingham	R.A.E.C.

GUARDS DEPOT
CATERHAM

"AT HOME"

SATURDAY
25th JUNE, 1955

The World Outside was Different

Of course it was. The men who lived and worked in the old Depot led an ordered, structured existence in a regime calculated to get optimum results from the training programme. But although the two were worlds apart, there was one aspect of barracks life which almost all civilian visitors commented on and would have taken back to their own world, had they been able — the standard of cleanliness throughout the Depot.

Outdoors, there was never any litter to be seen. Grass and verges were kept trimmed, trees and shrubs were pruned and pollarded, windows shone, paintwork of buildings was washed and renewed when necessary: signage was kept fresh and clear.

Few visitors, apart from those who attended the annual and very popular Open Days, saw much of the interiors. There, the standards were the same. All the floors in the living accommodation gleamed. All equipment, from furniture to fire-fighting apparatus was properly maintained. Stairs, entrances and emergency exits were clean and free from obstructions. Lavatories and ablutions, or washing areas, were similarly cleaned and inspected daily.

John Patrick Kenneally joined the Irish Guards in 1941 and won the Victoria Cross in the North African campaign of WW2. He did his induction training at Caterham and, as he says elsewhere in this book, 'There is not a hospital in Britain whose communal facilities could compare with the cleanliness of the Guards' Depot.'

The Grenadiers' archivist, Major Peter Lewis MBE, also began his career at the old Depot and recalled its cleanliness standards with not a little pride: 'You never saw debris of any kind anywhere. Nobody ever dropped anything — in autumn, even the leaves hesitated before they fell!'

And, unlike so many disfigured areas in the world outside, there were no graffiti artists in the Depot. Their feet wouldn't have touched the ground.

Contrary to the mischievous old stories, the practice of whitewashing coal was consigned to the annals of ancient history ages ago. But 'way back, this was sometimes done as an anti-theft precaution. It was not unknown at one time for men to take coal under cover of darkness from the bays where it was stockpiled. Giving the top layer a coat of whitewash meant that any overnight 'withdrawals' would be immediately obvious.

In the 1800s . . .

Getting the Adjutant out of his boots after a day's shooting at Marden Park.

The Birth of the SAS

David Stirling was one of the twentieth century's outstanding military figures. He was also one of the most unorthodox. Early in the North African campaign of WW2, he hand-picked a nucleus of fellow Guardsmen to create the ghostly units which operated far behind enemy lines, striking from the desert at German airfields, fuel dumps and communications - the SAS.

He also took part in the operations of G Patrol of the Long Range Desert Group, which consisted of volunteer Scots and Coldstream Guardsmen, and first went into action in 1940 led by Captain MMD Crichton-Stewart, Scots Guards. With the SAS, the LRDG did enormous damage to the enemy, destroying on the ground more German aircraft than were shot down in the desert war.

Stirling went through Caterham just before the outbreak of WW2 and was commissioned in the Scots Guards. Evidence of his spirit of adventurous initiative showed immediately. Russia, which had signed a pact with Germany, had attacked Finland in late 1939 and a Franco-British plan to aid the Finns centred on a joint force trained in ski and winter warfare, and which would reach them by crossing Norway and Sweden.

Because more senior officers with winter sports skills took precedence, the young Guards lieutenant resigned his commission and volunteered to serve as an NCO in what was to become the Scots Guards' 5th (Special Reserve) Battalion. Then he went with the Battalion to Chamonix to train with the Chasseurs Alpins, the French army's winter warfare experts. The Battalion was to be commanded by Colonel JS Coats, MC, Coldstream Guards, with Captain WDM Raeburn, Scots Guards, as adjutant - both winter sports experts.

The Battalion returned to Britain and was about to embark at Glasgow docks for Scandinavia when the Finns announced that they had signed an armistice with Russia. The expedition was cancelled, the Battalion disbanded and its members returned to their units.

Early in the North African campaign, Stirling badgered senior staff officers with his proposal to set up small, highly mobile and self-contained units which would take the war far into the enemy's rear positions and attack key targets. His persistence paid off and the Special Air Service - a meaningless name which would convey nothing of its operations to the Germans - was established. Its members were volunteers from the Guards regiments, selected for highly specialised training.

Stirling led the SAS till he was eventually captured on a raid. He escaped four times and was finally imprisoned in Colditz Castle.

LRDG and SAS played key roles in the desert, accelerating the defeat of the German Afrika Korps and enabling the Allies to take the war into the Continent of Europe.

Colonel Sir David Stirling DSO died in 1990 at the of 75. His SAS is still very much alive.

This photograph of David Stirling was taken in Chamonix after he had resigned his commission and was undergoing winter training in France for the expedition to Finland.

Lieutenant David Stirling, Scots Guards (8th from left, front row) at a Scouting, Observation and Sniping course at Bisley in 1939. During the North African campaign of WW2, he persuaded senior officers to let him set up the SAS, which immediately proved itself in operations behind enemy lines.

The Missing Pictures

Sadly, very few photographs exist which show what the Depot looked like in WW2, the period when thousands of men were constantly under training in a hugely expanded area and the threat of enemy action was real and immediate. Whereas from WW1, there are at least some photographic records of what the place looked like, the visual evidence of the large wooden hut complex, the emergency water tanks for use against incendiary bombs, and the additional drill squares of the 1939-45 period, is lacking. The squad passing-out photographs continued to be taken, but the ever present threat of spies and 'fifth column' infiltrators resulted in far tighter security than in WW1 and severely regulated all other photography. In most cases, it was banned outright.

In the 1800s

A Caterham parade guard of honour in July 1887. No sign of wilting despite a shade temperature of 90° according to the anonymous artist. The VIP being honoured must have been a big noise indeed!

Recruits and Guardsmen

They're changing the Guard at Buckingham Palace
Christopher Robin went down with Alice;
Alice is marrying one of the Guard,
'A soldier's life is terrible hard',
Says Alice.

The Tallest Soldier in the British Army
Copyright
1. Pte. H. Barter. (of Totnes. Devon) Age. 18. Height. 6 ft 8½ in.

A Family Tradition

The contribution of the Guards to the security of Britain and the defence of her former Empire runs like an unbroken thread through this country's historical tapestry. Within it, across the centuries, family names recur. Men have followed fathers, uncles, grandfathers into the three older regiments — Grenadiers, Coldstream and Scots Guards. And there is every indication that the same family tradition is being followed in the Irish and Welsh Guards, which were both established this century.

This tradition of continuity is illustrated dramatically by the Rigley family. During the dangerous decade spanning WW2, four Rigley brothers did their initial training at the Caterham Depot, one after the other, three going on to serve in the Grenadiers, the fourth in the Coldstream. Only one brother survives to-day. Victor Rigley, who served in The King's Company of the Grenadiers' 1st Battalion, tells his family story.

"Looking back, I share the mixed feeling of nostalgia and pride of men who passed through the awesome gates of the Caterham Depot. I joined the Grenadiers there in November, 1938, transferring from the Sherwood Foresters with the number 4975210 to follow my older brother, George, who had already passed through Caterham and was 2615624 Corporal Rigley G with the 2nd Battalion at Wellington Barracks.

Shortly after the outbreak of war in 1939, my younger brother, Frank, joined the Coldstream (2665934). And at the end of the war, my brother, Roy, arrived at Caterham to become a Grenadier recruit with the number 22213344. He went on to become a Sergeant in the Signals Department.

One memory stands out — my first guard duty at the Depot. I was raw, and because the Sergeant of the Guard was Irish I had difficulty understanding his orders and words of command. In short order, I was on a couple of charges for not obeying his instructions properly.

I was on midnight till 2am sentry duty opposite the guardroom. I stayed as alert as I could, although nothing happened to break the tedium. When the Depot clock eventually struck 2am, I heaved a sigh of relief and waited for my replacement to arrive. Nothing happened. When the clock struck 2.15, I sensed something was amiss. I knew that leaving one's post on sentry duty without being relieved by an NCO was a serious offence. But I also knew that the Sergeant already had me on two charges, so I thought, what the hell, and went across to the guardroom.

Everyone was asleep.

When I woke the Sergeant, he almost had a fit. He told me to stay in the guardroom and posted my relief right away. Then he roused the Corporal of the Guard and rushed him out to relieve the other posts. Although I was still worrying about my forthcoming punishment, I climbed into a bunk and slept. Morning came and the guard was dismounted and dismissed. As I was about to leave, the Sergeant took me aside and said, 'ahem!' that he thought the two outstanding charges could be overlooked this time. Strange, I had no difficulty then with his Irish brogue — or his reasoning.

During the Winter of 1938, two fighter 'planes collided over Caterham during night exercises and the wing of one fell next to the barrack block I was in. Rumours of war were rife at that time, and many of us thought at first that this was the start of hostilities. It sounded like a bomb.

We had a Colour Sergeant, 'Timber' Wood who was big, even to us. In the jargon of the day, he was 'dead regimental', a stickler for all the detail of tradition and behaviour. He would make evening spot checks in the barrack rooms, bawling 'Stand by Your Beds!' Then he would inspect each man, tugging at their braces to test for loose buttons, and almost lifting them off the floor. At the same time, he would be firing off questions about regimental history and battle honours.

The Depot gymnasium was known as the 'torture house'. Press-ups and hanging from the wall bars with legs outstretched was bad enough, but worst of all was rope climbing. If you got stuck half way, the gym instructor would take the slack at the rope bottom and whack your legs with it to encourage you to get to the top.

As our squad instructor, Sergeant Holding was good but very tough. He would dismiss us from drill parade at the entrance to our barrack block, and we would dash to our room on the top floor to change for PT in two minutes flat, even although we wore puttees at that time. It seemed that almost before we reached the room, he was yelling, 'On Parade'. It was Winter and a bucket of soup or a hot drink had been placed on the top landing for us. We never got any. Sergeant Holding regarded it as 'time wasting'.

The first week we were squadded, he marched us over to the NAAFI and our hearts rose at the prospect of 'tea and a wad'. He lined us up at the counter and told the NAAFI girl: 'Each man wants two dusters, a tin of Bluebell, a tin of Kiwi black, and a block of blanco'. And we had to pay from the eight shillings a week we were left with, after stoppages for barrack damages and enforced savings to fund us when we got our first leave, etc. No tea, no wads . . .

I particularly remember Recruit G Webb who is in my squad photo. He was the same age, and we were confirmed on the same Sunday in the Depot chapel. He was killed in June, 1940, aged 20. I have a photograph of his grave in a village on the border of France and Belgium.

The Rigley family have pride and a sense of achievement in the service my brothers and I gave in the Guards Brigade. George served with the Grenadiers in Germany and Holland and saw action at Arnhem. Later, he was an instructor at Windsor Barracks; Frank served in Italy with the Coldstream; Roy in North Africa with Grenadiers; and I was at Dunkirk and served in NW Europe with the Grenadiers' 1st Battalion.

The others are gone now, but I think I speak for all four of us ..."

The Man on Hill 212

The 1st Battalion, Irish Guards took and then held Hill 212 against repeated German attacks in the latter stages of the North African campaign in WW2. It was brutal, hand-to-hand combat between battle-hardened adversaries, and turning back the waves of Wehrmacht assaults played a key part in the Allied victory, which culminated in the fall of Tunis. The Hill 212 action was won at huge cost to the Micks. Of the battalion which went up the hill, only 80 men came down it at the end.

One of the men who did was Lance-Corporal John Patrick Kenneally, whose military background would strain credulity in a work of fiction. An Artillery soldier at the outbreak of war, he was rejected for active service as too young when his unit went to France. Then he was drawn into a pub fight, lost the cap which would identify him while escaping the police, and decided to desert. He joined the gang of Irish workmen involved in the pub fracas and travelled the country with them, blacking-out buildings and working on bomb damage. At the end of a month when the Irishmen were returning to Ireland, one of them gave him an Irish ID card.

At that point, the man who had been born Leslie Jackson in Blackpool, became John Patrick Kenneally. Under that name, he immediately joined the Irish Guards, a regiment he had long admired. And it was under that name on Hill 212 that he won the Victoria Cross.

Ordered to a forward position to warn of German attack preparations, he saw the enemy in company strength below him, being briefed by an officer. Single-handed, he charged down the bare hillside, firing his Bren gun from the hip. The Germans broke in disorder and Kenneally returned to his observation post, from which he harassed their retreat.

Two days later, accompanied by a Reconnaissance Corps sergeant, he again charged an enemy company forming up for an assault. With his Bren, he inflicted many casualties and broke up the attack before it could start. Although wounded, he refused to give up his Bren and continued in the defence of Hill 212 throughout that day.

In Tunis, Kenneally received the medal from General Alexander, the 1st Army Commander and an Irish Guardsman himself.

To-day, at his Worcestershire home, Kenneally remembers Caterham with a touch of pride.

"Caterham had an ethos all its own. The Depot buildings and their facilities matched the turnout expected of the Guardsmen. They were always immaculate — even the ablutions and latrines. There is not a hospital in Britain whose communal facilities could compare with the cleanliness of the Guards' Depot.

Ages in my squad ranged from 18 to 32, and Sergeant Shaun Carr had 17 weeks to turn this motley crew into Irish Guardsmen. First, he taught us how to stand, and then how to walk. In the drill sheds beside the square (recruits were not allowed on the square till they had attained some sense of order and were able to march together) we stood against the wall with heads, shoulders, backsides and heels pressing against it. This is not as easy as it looks, and was often used as a punishment for slacking. After some time, this training gives a man bearing and forces him to hold his head up.

Anyone can walk, you may say, but there is the right way, the Guards' way; heel-and-toe, heel-and-toe was drummed into us. We did many foot exercises to get our feet and ankles lithe and supple. We did two one-hour periods of PT every day. I had done plenty of PT in my time, but this was something else. Their aim was 101 per cent physical fitness.

We were going to need it for the hours and hours we would spend on the barrack square.

The rank of Trained Soldier is unique to the Guards. He is a Guardsman but wears a five-pointed star on his right sleeve. In my day, it bore the motto of the five Regiments of Foot Guards, the Guards Brigade, '*Quinque Juncta in Uno*' — Five Joined in One. It reflected the bond of comradeship binding the Bill Browns, Lilywhites, Jocks, Micks and Taffs.

The Trained Soldier was usually an older man and was a father figure to the young recruits. It was he who taught us how to lay out our kit, how to polish buttons and boots, to fold blankets to a millimetre of accuracy, and how to lay the rough battledress trousers down, sleep on them and produce immaculate creases. He taught us how to scrub, clean and polish the barrack room furniture. The fire buckets and stove shone like silver. The Trained Soldier answered all our questions. He told us what to do and what not to do, and pointed out the many and varied pitfalls we were bound to fall into.

Like the Squad Sergeant, the Trained Soldier was a linchpin in the Guards' training regime...

We pounded the square and gradually, very gradually, we began to improve. As we got fitter and learned the drill movements, it got a little easier. Being pitch-forked into this crazy life had been a severe culture shock for some of my more sensitive comrades. There was no privacy whatever; we ate and slept together, and we bathed together in the communal baths. It was no place to be shy or retiring. I noticed that the rougher men from poorer backgrounds (and I was one of them) took to Army life much better than their comrades from respectable middle-class homes. If you had left little of any consequence behind you, this hard life was easier to take.

We were very fortunate to have Sergeant Carr as our instructor. He went by the book and was very strict but scrupulously fair. If a man was flagging, he didn't bawl and shout at him but took him aside and tried to encourage him. But if we were slack or idle, he would throw the book at us. I developed a great respect for him. To me, he was the epitome of what an Irish Guards non-commissioned officer should be. He had served in Egypt and Norway and was to suffer terrible facial wounds in North Africa, which finished his career as a soldier. I am certain that if he had come through the war unscathed, he would have achieved high rank in the Regiment. I have never forgotten him. He was always an example to me.

The day of the passing-out parade eventually arrived and as we formed up with identical squads from the other four Regiments of Foot Guards, Sergeant Carr impressed on us that we had to be the best and a credit to him. It was like a miniature Trooping the Colour. In review order, we were inspected by the Commandant, followed by the Adjutant, Company Commanders, the bristling Depot RSM, the CSM, and Uncle Tom Cobley and all. We marched past the Commandant in quick and slow time to each regimental tune. Then we did it again in review order.

Military music never fails to get one going. The adrenalin flowed and we dug our heels in and swung our arms with the best of them. It was great to be a Guardsman, and we didn't let Sergeant Carr down.

Next day, immaculate as new pins, we formed up outside Company HQ before leaving Caterham for the Training Battalion. The Company Commander told us we had done well and wished us luck. We piled into the transport, Sergeant Carr gave us a nod, and we were away on the next stage of our adventure. As we drove down Caterham Hill, we passed some young men struggling up it with their cases.

Poor sods, we thought."

In March 1998, Queen Elizabeth the Queen Mother, was recovering from a hip operation at the age of 97, and wasn't able to present the traditional shamrocks to the Irish Guards on St. Patrick's Day. Called on to deputise for her was the regiment's former Sergeant, John Patrick Kenneally VC. The photograph shows him after he made the presentation at Wellington Barracks.

The autobiography, Kenneally VC, is published by Kenwood, Cringle Cottage, Crangle Field, Stockmore, Huddersfield HD4 6YE.

In the 1800s

Performing Cane Drill

The Drill Expert

'My Goodness,' exclaimed Her Majesty, 'that was impressive!'

The scene, Buckingham Palace. The occasion, the Sovereign's inspection of the elite Queen's Company of the Grenadier Guards. Responsible for preparing the Company for the ceremony, RSM Peter Lewis found that he also had a problem, imposed by the restricted space available. Drill manuals didn't provide a solution.

So Lewis devised a specially compact drill movement which enabled the company to form from column of divisions into line, and simultaneously from three ranks into two for the march-past. Complicated as well as new, the movement had to be practised until Lewis was satisfied.

The Queen's comment was the result.

Long recognised throughout the Guards Division as an outstanding expert in drill, Lewis achieved his high standards on the square and in ceremonials because he got so much satisfaction and personal pleasure from that attainment.

Few could equal his expertise in this area. He could drill two squads at the same time, one slow marching, the other marching in quick time — and with his back turned to both squads!

Peter Lewis began his Guards career at Caterham Depot in 1945, where RSM S M Hamilton MBE, Scots Guards, was heading the staff.

Lewis began to learn the breadth and detail of his chosen career under Hamilton, a process he was to continue at each stage throughout his service.

By 1947, he was a Lance-Sergeant with service in Palestine and Malaya on his record.

Regarded by many as the very embodiment of a Guards RSM, Grenadier Peter A. Lewis was a product of the old Depot. His code was discipline with humanity — a policy that yielded outstanding results..

He was promoted Gold Sergeant in 1951 and achieved an 'A Outstanding' grade at the School of Infantry Support Weapons Wing, Netheravon.

In the Army Education Exams he was one of only two to pass in London district. Armed with this distinction, he progressed to CQMS and Pay Sergeant of Support Company. Promotion to CSM came after postings to North Africa, Egypt and Cyprus, and led on to a tour on the staff at the Royal Military Academy, Sandhurst, under RSM Lord.

His next appointment took him back to the Guards Depot, where he put Lord's methods to work. Two of his achievements were cause and effect — he raised already high standards among recruits undergoing training, thereby helping to cut drastically the numbers of recruits seeking to 'buy themselves out'.

Lewis had the ability to cope with even the outlandish amongst the unexpected. Training the Royal Company of Archers, the Queen's Bodyguard in Scotland, for a parade, he devised a new form of bow drill to suit the occasion. One nationalist-minded Archer asked him why the task had fallen to a Grenadier and not to a Scots Guardsman. Lewis gave a typically direct answer: 'Gentlemen, this is an important parade for Her Majesty. You *need* the best — and I *am* the best!' Those who knew him as a Drill Sergeant with the Grenadiers' 1st Battalion, and later as the same battalion's Regimental Sergeant Major, would agree.

When he was commissioned Captain Quartermaster of the 2nd Battalion, Colonel David Gordon-Lennox assessed him thus: 'He has always made an outstanding success of anything he has done, because he has that ability, determination, and sense of loyalty and responsibility which are given only to those who reach the highest levels of leadership.

The regiment will always be grateful for what he has achieved.'

When he retired in 1977, Peter Lewis summarised Caterham's legacy thus:

> "We learned many things at Caterham.
> "We learned the importance of punctuality — always being where you were meant to be five minutes before the time required.
> "We learned to appreciate the beauty of ceremonial and the precision required to achieve a good result.
> "We learned that whatever we do, to do it properly — the Guards' discipline that prevails, whatever the circumstances.
> "Perhaps many of today's social problems would be solved if only all young men could experience 16 weeks of that training."

Peter Lewis retired with the rank of Major and continued to work as his regiment's archivist at Wellington Barracks. Although ill, he continued working till within a few days of his death in 1998.

————oOo————

Guards' Humour

Caterham Depot: 'It's the only place in the world which looks like a prison, where you're treated like a criminal, and yet you spend the rest of your life telling everyone how much you enjoyed it!' - Major PE Lewis, MBE, Grenadier Guards.

————oOo————

A Service Secret

'Sixteen!' The recruiting sergeant at Forres in the north of Scotland laughed. 'Ye'll no get into the Guards at sixteen, laddie.' He dropped a heavy wink. 'Come back in a fortnight when ye're eighteen.' Two weeks later, Bill Shand was on his way south to Caterham — signed-up, King's Shilling and all, a recruit in the Scots Guards. Throughout his long service, his secret was never discovered. It was a cold morning in January, 1938 when Bill arrived at the Depot and the details remained etched on his memory with sharp, wintry clarity. The Coldstream sentry, the immaculate Corporal of the Guard, a first sight of the drill square with hundreds of recruits still in civilian clothes being pursued by bawling instructors. And then his communication problems when the Doric Scots of his Moray fishing village collided with the thick Irish brogue of his Trained Soldier.

"I had enlisted in the Scots Guards, so it was just one of a series of shocks to find an Irishman in charge of the hut where I and several other Scots recruits were given bedspaces. The bugler's reveille wasn't enough for him. Each morning, he ran up and down the row of beds, hammering at a dustbin lid with a soup ladle. Fortunately, we were soon moved on to Victoria Block, 'K' Company, Scots Guards, and Trained Soldier Rayson, a six-foot-four Cumbrian who introduced us to discipline and the relentless regime which would make Guardsmen of us. The continuous, day-long training programme gradually wore away our civilian awkwardness and set the beginnings of our squad camaraderie which became stronger with each day. It was hard. We went to bed exhausted and reveille came too soon. But we were well fed and fitter than we had ever been. And as we began to perform like soldiers, the hectic schedule began to seem like second nature. The system had stripped us down to the bare essentials.

After about nine weeks, we were an audience one Saturday at the main square where Guardsmen who had finished their training were assembled in red tunics and bearskins. We watched as they were put through a drill programme by the RSM - all of them slick and professional. It was a shrewd psychological move and impressed us deeply. But I'm sure we recruits all wondered if we could ever reach that standard ourselves.

I remember a lesson in observation given me by Sergeant Smith, the musketry instructor. Out on the playing fields one day, he told me to double over to the end of the nearest building and see if I could see the Depot clock. When I returned and said I could, he asked what time it was. I hadn't noticed, so I had to double back to find out. When I reported, he barked: 'Now tell the squad what kind of figure four is on the clockface.' Again, I didn't know so I had to do the round trip again at the double. We were in full field service order, complete with rifle, and the lesson went home. Even to-day, sixty years on, I always notice whether the four on a clockface is IV or IIII.

We had been miners, dockers, mill workers, city slickers and farm workers, speaking in all the dialects of the UK. The range of intellect and education was also wide — from a university drop-out to another who had difficulty writing his name. Before we left the Depot, he could write home like the rest of us. Our departure was very different to our arrival. We marched down the hill to the station behind two pipers, a well-drilled squad. People stopped to watch, some even clapped their hands. It was a great feeling. Behind us were the men who, in six months, had knocked us into shape and left an indelible impression, the grip of which lasted a lifetime.

At Chelsea Barracks, we were dispersed to different companies and then fitted out with tunics and bearskins. There followed two weeks of intensive drill to smooth out our remaining recruit practices, particularly in the art of precision double sentry duty.

Finally, the big day arrived. I was in the 2nd Division of the King's Guard at the changing of the guard at Buckingham Palace — probably the proudest day of my life. I was a Scots Guardsman at last. For a time, I missed the rolling Surrey countryside and the towns of Croydon, Purley and Coulsdon and Caterham Depot. But it was June, 1938, I was 16 years of age, and a summer of London duties lay ahead."

And WW2 was just over the horizon ...

Bill Shand went back to his Moray village at the end of his service. He died in 1998.

—————oOo—————

Guards' Humour

How distant an invisible enemy rifleman is can be judged by the crack-and-thump principle, the young Guardsman is taught. The crack of the supersonic bullet passing (hopefully) is followed by the thump of the distant shot. The time interval between the two allows a rough estimate of the range involved. The two Scots Guards weapons instructor sergeants at Caterham in the 1940s - one slight, the other heavily built - were always referred to as Crack & Thump.

—————oOo—————

You Want to WHAT??!!

Every Guardsman who ever did his time at the Caterham Depot fervently wished at some time that he was somewhere else. The wish, however, can seldom have caused the shock wave which went through the barracks in 1936, not long after Steve Wheeler signed on in the Coldstream Guards.

His whole squad applied to transfer to the Navy!

The Navy didn't get them, and Steve soldiered on in the Coldstream. Nine years later, 2657401 Sergeant Stephen Keedwell Wheeler received the Military Medal from Field Marshal Montgomery for his part in a tank battle in Germany.

"The idea of joining the Navy instead came about because we thought we were having a rough time. To us, the change from Civvy Street seemed harsh as well as abrupt. When we arrived as recruits, a towering Irish Guardsman met us in the receiving room, handed each of us a piece of blue mottled soap and barked: BATH!!

We soon learned that cleanliness applied to every aspect of our life at Caterham. Codrington Block where we were based was spotless. The iron banister rails on the stairs upper floors were burnished to a chromium-like finish, the barrack room shone, and woe-betide any recruit who made the slightest mess. The contrast shock of this sudden shift from civilian to Depot life led to the whole squad putting in for a transfer to the Navy. The application was organised by the one recruit among us who should have known better.

He had been to Caterham once before, but when it was discovered that he had joined under age, he was told to come back at 18 — and he had.

Word of our attempt to get out went round the barracks like wildfire. A whole squad being marched in on Company Orders was unusual to say the least and there was a lot of speculation about what was going to happen to us — apart from not going to the Navy. In the event, it was a very brief hearing and we were returned to duty pretty sharpish!

Strangely, it was a turning point for the squad and under the round-the-clock supervision of our instructor, Lance Corporal Skells and Trained Soldier Littlehales, we put our imagined grievances behind us and made good progress in our training. We passed all our periodic and increasingly difficult inspections at the first go and left the Depot with the right to call ourselves Guardsmen.

A year later, I was back at Caterham as a Trained Soldier and officer's servant. This time, I made two bad mistakes. The first was to complain to the Duty Officer about the meals. The second was being caught in bed half-an-hour after reveille.

For the first, I had to attend all the mess meetings till further orders. For the second, I was on Adjutant's Orders and had to 'answer my name' to the Sergeant of the day Guard at reveille for three weeks.

Enforced attendance at those mess meetings showed me the humanity underlying the hard discipline. At one, I suggested that treacle tart would be a nice change from plum duff. It was turned down.

About a week later, the Duty Officer arrived at the dining room at lunchtime and handed me a parcel. Not wanting to open it before the recruits at the table, I waited till I was on my own.

Inside — a treacle tart! Major SBR Cooper, the Quartermaster, who attended all the mess meetings, had got his wife to make one for me.

Even being marched in on Adjutant's Orders had its funny side. Five of us had been caught oversleeping after reveille and, as we were from four of the five regiments, we went in according to the traditions of our regiments — one at the salute, one with no cap on, one saying 'thank you for permission to speak', and myself with arms held at my sides. Even the Adjutant had to smile when he saw us.

I did about three years as an officer's servant, including a stint in reinforcements to the British Expeditionary Force in France.

I got back to England in June, 1940, and joined the 4th Battalion. There, I felt it was time to be a real soldier again and managed to get back to duty.

At Boreham Wood where we were stationed, the officer I had worked for put me in for my Lance Corporal tapes. I didn't want them because my drill was rusty and new rules had come in about forming threes instead of fours, and I had never done this.

Left with no choice, I did a drill course and was sent on a Corporals' Instruction Course at Caterham. Every regiment in the Army was there in a move to get them all to conform to the same method of drill.

We were treated just like recruits, with a squad instructor and trained soldier — marched to baths and haircuts, the only difference being that we were allowed out at night. In passing out from the course, we took turns drilling the squad before the inspecting officer. The report I got was better than I had expected.

At Caterham, we were told to lay our kit out for inspection, and we made a very good job of it. 'Not good enough', said the inspecting officer, adding: 'I'll see it again at 1800hrs.' This didn't go down too well with us, so we just left everything as it was. 'That was a lot better', said the officer at 1800hrs.

Coldstream Sergeant Steve Wheeler received the Military Medal from Field Marshal Viscount Montgomery.

When the 4th Battalion got tanks, I did a tank commanders' course at Bovington and took over a Churchill tank in No 2 Squadron. In July, 1944, when the Brigade landed in Normandy, I was a Full Sergeant. Because of casualties, I was put in charge of a troop of three tanks.

The citation for his Military Medal takes up the story . . .

THE MILITARY MEDAL.

No.2657401. Sergeant Stephen Keedwell Wheeler.

COLDSTREAM GUARDS.

On the afternoon of the 28th.March 1945, during the advance on MUNSTER, the leading Squadron was held up some 3 Km to the East of Dorsten by at least two Batteries of antiaircraft 88 mm Guns and a quantity of 20mm.

No.2 Squadron 4th.Battalion Coldstream Guards was ordered to outflank the enemy position, Sergeant Wheeler commanded the leading Troop of the Squadron in this manoeuvre. He first located and destroyed one of the guns and then with great skill so manoeuvred his Troop that he reached the high ground without loss. From this position he dominated the enemy's gun position, and by his accurate fire forced the crews to abandon their guns.

Sergeant Wheeler displayed outstanding leadership. His sound appreciation and his skilful handling of his Troop which resulted in the destruction of the enemy guns, allowed the advance to continue and valuable ground to be gained...

In the 1800s

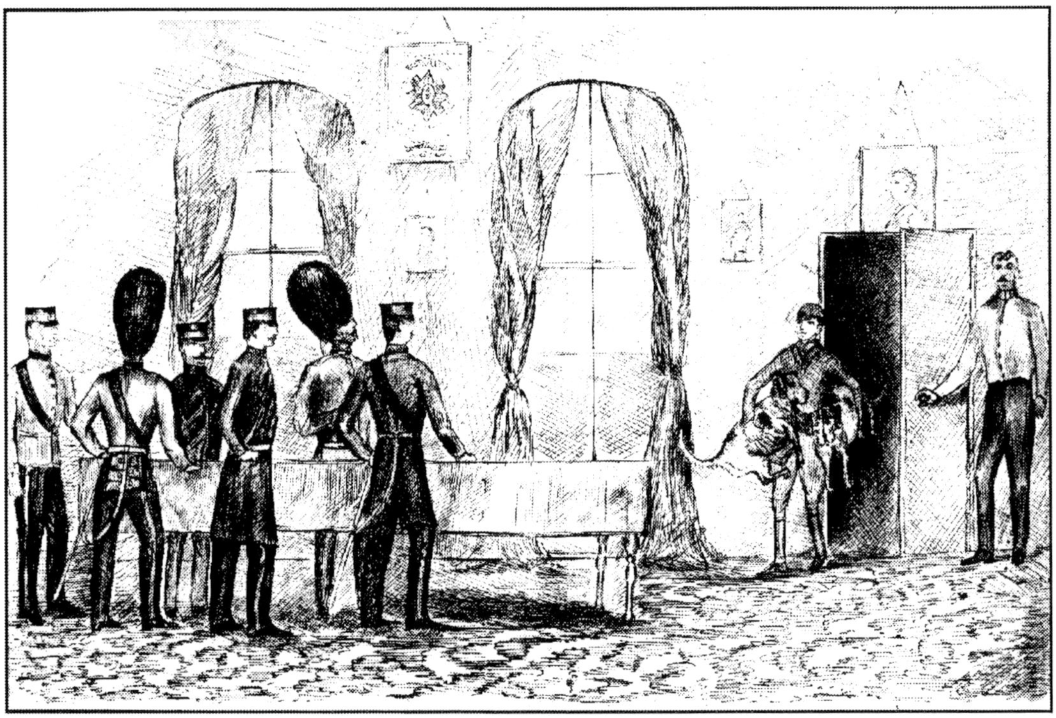

Many of the officers rode to hounds in the 1800s, and a pack of foxhounds was kept in the Depot kennels. This anonymous artist pulled the legs of the huntsmen by suggesting they took the sport seriously enough to have each puppy 'intake' inspected in the mess!

Caterham, Beirut and Self-Respect

Terry Waite survived five years in an underground Beirut prison cell, four of them in solitary confinement. For most of that time, the world didn't know whether he was alive or dead.
One of the things which enabled him to survive that ordeal was his experience of forty years earlier when he joined the Grenadier Guards at the Caterham Depot . .

"At the age of 15, I was bored with school and anxious to see the world. I pleaded with my father to allow me to join the Navy. He talked me out of it. Two years later, I decided that as National Service was approaching and would require two years of my life, I might as well sign on for three and receive a better wage. Being six foot seven, I thought of the Guards.

'You do realise,' said my father, 'that Caterham is probably the toughest place in the British Army.'

'Yes', I replied, thinking I knew what was meant by 'tough'.

I made my way to the recruiting office in Liverpool, collected my rail warrant and was put in charge of two other National Service men.

'Get there by early afternoon,' said the recruiting sergeant. 'Don't hang about in London.' The three of us walked to Lime Street Station.

I was the only one who had previously travelled to London, which probably explained why I was put in charge of the party.

Clutching our cardboard suitcases, we arrived at the main gate of the Guards' Depot, Caterham.

'Stand up straight, lad', bellowed the duty sergeant as I handed him our papers.

'Yes, sir'.

'Don't call me sir', he shouted. 'Sergeant, understand? Sergeant!'

It started to rain.

'That way', he bellowed, pointing towards the Depot. We started to walk. A voice boomed behind us, 'At the double. At the bloody double!'

We quickened our step.

'Are you deaf as well as stupid?' thundered the voice. 'At the **DOUBLE!!**'

We trotted down the main drive, past the chapel alongside the parade ground and into the barrack block. We were in the Guards and life would never be the same again.

I can remember my days at Caterham as though they were yesterday. The miserable state of my kit as I collected it from the stores. Boots that had to be burnt off with the back of a spoon heated over a candle. Shuffling on to the parade ground so that the thick layer of polish wouldn't crack before we were inspected. Running through the local woods in the height of summer, wearing a gas mask. The indescribable mess of used pots and pans which confronted me when on fatigues in the kitchen. The despair when we returned to our block on the second floor to discover that all our carefully folded kit had been thrown through the window by an irate Trained Soldier.

And yet, almost forty years later when I was sitting deep underground in a Beirut prison cell, I remembered the hard training I had received. During the first weeks of captivity when I still had my own clothes, I folded my trousers at night and placed them under my mattress on the floor. I asked my captors for shoe polish. They thought I was mad but I knew, thanks to Caterham, that it was essential to keep my self-respect.

When I was finally released some five years later, after spending four years in total solitary confinement, I was taken to Syrian Intelligence Headquarters. 'What would you like,' inquired an official. 'A haircut,' I replied.

I was determined to show the world I had not been defeated and would go out on public parade as well turned out as possible.

Caterham? I can't say I loved the place, but it taught me valuable lessons. In fact, it certainly made a major contribution to my own survival. I've never gone back following my discharge, and now the Depot has moved.

However, it lives on in my memory. My father was right. It was tough. But unknown to me, life was to deal me a tough hand and I shall always be grateful for what I learned, 'way back in the distant '50s.'

Guardsman Waite — ' a very junior member of a very senior regiment'.

The Guardsman who Sparkled

WB (Bill) Davies joined the Welsh Guards at Caterham on Man Service in June, 1939. In reality, he was just two months short of his seventeenth birthday, but his youthful ambitions were already fixed on a career, leading to the legendary field marshal's baton. He didn't make it. But who can say he wouldn't have succeeded if a badly damaged right arm hadn't ended his Guards career. His progress up the promotion ladder had been rocket-like: Lance Corporal and Squad Instructor at 17; Lance Sergeant and then Sergeant at 18; Platoon Commander of No 3 Platoon, No 1 Company, 3rd Battalion Welsh Guards at 20, serving throughout the North African campaign of WW2. Two months short of his twenty-first birthday, he was CQMS of Support and HQ Companies. Less than two months later he became CSM of No 1 Company, and finally, he was appointed Junior Drill Sergeant of his regiment's 3rd Battalion at the age of 22.

The vivid memories Bill Davies has of the old Depot are in two time-frames; first, his recruit induction and then his return as a squad instructor. A copybook Guardsman he may have been, but life at Caterham taught him anyone can come badly unstuck . . .

"After 12 weeks squadded, I was selected to be Commandant's Orderly to Lt-Colonel A L Cassenove MC as from the following Saturday. For days (and nights) I buffed and pressed and polished. Then I was inspected by Trained Soldier Shackles, Lance Corporal Griffiths, Sergeant Harry Randall, CSM Jones and CSM Fitzgerald, before reporting to the Officers' Mess. When the Commandant appeared, I went into the well-rehearsed ritual: '4034108 Recruit Guardsman Davies, W, Lance Corporal Griffths' squad, No 8 Company, Welsh Guards reporting to you as your Orderly, Sir!' I was inspected by The Great One and after answering a few questions, escorted him to the previously detailed areas and places of his programme, finishing up outside his office.

RSM McKinney, Irish Guards, was there. The Commandant spoke to him and the RSM then addressed me in the dulcet tones peculiar to his rank — a cross between the bellow of an angry bull and the rasp of a circular saw. I was a bundle of hypertension. Then suddenly, he said in a surprisingly mellow voice, 'Well done, lad. Report to your CSM and tell him you have been awarded two credits by the Commandant, one for turnout and one for your personal drill'.

On cloud nine, I reported to Edward Block where the Company was readying itself for the Saturday morning Commandant's Parade. I told Griffiths about the credits, he in turn told the CSM, and seconds later Major Rice, the Company Commander, was adding his congratulations. Then he told the squad that he would arrange a special tea for them — fish and chips, pineapple chunks and custard!

I really don't know to this day how the next incident occurred, but it was horrific.

As Commandant's Orderly, I turned out for the big parade in buff belt and bayonet, with buff pouch slung over the shoulder and resting between my shoulder blades. In the rush and bustle, I was fitted out by eager hands, brasses rubbed, serge trousers brushed, boots flicked, and ears ringing with repeated commands to 'get a heave on!'

We formed up on the perimeter of the parade ground, standing easy, when a sibilant voice filled with a mixture of fear, hate, disbelief and venom hissed the words which shattered my day: '08 Davies, where is your bloody bayonet?' I checked with a furtive hand and found . . . nothing! Instead, I felt a bottomless void in my stomach and a moist feeling in my drawers cellular. A banshee-like scream from the RSM had me and a dozen other right-hand men performing the fixing of bayonets. Although bayonet-less, I went through pantomime motions of fixing one.

We were inspected by the Company Officer, Mr Ricketts. Then we marched past the saluting base in both quick and slow time and — incredibly — I rubbed! ! Had I got away with it, thanks to my continuous prayers to my Maker as I marched? Prayers like, 'Please, God, don't let me get booked and I promise to attend voluntary Bible readings every week if I get away with it .. . ' I don't know if it was divine intervention or temporary blindness in the powers that be, but get away with it I did. Later the Black Prince (as we called Corporal Griffiths) gave me a tongue-lashing, promising me a feed of a very different nature to the fish and chips Major Rice had promised the squad.

It was said that when TS Shackles found my gleaming bayonet and its virgin white frog propped against a wall, he thought firstly about desertion, followed in quick succession by suicide, or a plea of temporary insanity. Instead, his revenge was one of many practised by Guards' Depot Trained Soldiers. For weeks afterwards, I performed every lousy chore known to mankind.

Prior to our squad passing out, WW2 was declared and I found myself on 'barrack stag', doing the first wartime sentry duty on the Depot's back gate, complete with a bandolier of ball ammo and a loaded magazine in my rifle.

I was Orderly Room Piquet Sentry when I dropped my next clanger. I had to deliver a note to Coldstream CSM Middlestitch at Codringham Block. Arms flailing, chucking salutes around like confetti at a wedding, I skidded into the block quadrangle and sighted two Coldstream CSMs. One, I knew to be CSM Nunn, so the other surely had to be CSM Middlestitch. I halted before them and after slamming my left tab through the cobblestones right up to the third fold of my putteed leg, asked permission to speak. Permission granted, I stated: 'A note for CSM Middlestitch, Coldstream Guards, Sir!'

Suddenly, the sun lost its warmth. An icy coldness stole over me from the Coldstream. The massive head turned my way, revealing a deeply-tanned face, vivid blue eyes, and a mouthful of gleaming teeth framed by a luxuriously curled moustache. 'WHAT?' The single word hit me like a cricket ball.

I repeated my message and received a double 'WHAT? WHAT?' like blasts from a twelve-bore. I felt I'd better change my style of delivery, so taking a deep breath, I said: 'I am the Depot Orderly Room Piquet Sentry. The Orderly Room Sergeant has given me a letter to deliver to CSM Middlestitch — are you he, Sir?' Before my eyes, he seemed to swell and grow. His face darkened and the white teeth were bared in a wolf-like snarl. I was able to observe these features closely because by now they were merely inches from my pallid face. Then came the word — 'MiddleDITCH!!!' It hammered at my eardrums and went on vibrating in my head as the CSM read his note. Timorously, I asked leave to fall out. This was granted, and after a number of as-you-were's, I was eventually able to make my escape.

I stopped to recover from my jellied legs at the porch of the block. Next minute, a 'Lilywhite' Trained Soldier bawled my ass off, ordering me to move off in quick time and not defile the sacred 'Coldie' brickwork with my filthy Welshness. One of my ambitions took on an extra intensity that day. One day, I would be a CSM and I would take a savage delight in chewing the ass off everyone inferior to me, and boy would they be inferior — especially Coldstream!

We passed out and I found myself at Roman Camp, Colchester. On parade, people avoided me and my sparkle. And it was the sparkle which had me detailed as demonstration man for the officer cadre course run by the then Major Greenacre, MVO. I demonstrated how a Guardsman should appear for all types of parade and duties. This resulted in me being allowed to drill a squad under the eagle eyes of RSM 'Micky' Dunn and Drill Sergeant Miles Davies. A month later, I was back at the Caterham Depot along with Lance Corporals 'Foggy' Farr, Eddie Stretton, 'Mucker' Miles, '35 Davies, and others my failing memory cells can't recall, facing the unknown perils of becoming possible squad instructors.

That weekend, we 'bobbed' away at our kit. Monday found us in a Brigade Drill Squad, ready for inspection by Captain Brownlow, the Grenadier Adjutant. I must have caught him on an upstroke because he told the accompanying sergeant, 'Give this Welsh Guards corporal a credit for his chinstrap.' The resulting rosy glow was short-lived. Behind me, I heard a rapid tattoo of tab slamming and THAT voice asking, 'leave to join the parade, Sir!' Then the voice went on to address the squad: 'When I come to you, state your number, rank, name and company.' He appeared before me and I stated my particulars. It was the same tanned face, blue eyes and white teeth, framed by a luxurious moustache — now Drill Sergeant Middleditch. My details were entered in his book, he passed on, stopped and then moved back to me. 'I know you, don't I, Lance Corporal!'

I heard my answer, not in a voice you would associate with a potential squad instructor — rather that of a stricken person, devoid of all hope. As a result, the remainder of that week found me showing some item of kit at Adjutant's Memo, courtesy of the Drill Sergeant.

Friday morning, I was called out to drill and instruct the squad. I gave my all. My commands were sharp, loud and clear and any error was corrected with a caustic edge. There was a tap on my shoulder and THE VOICE said, 'Halt the squad, stand them easy, report to the Adjutant.' I marched up to the Adjutant, cracked the tarmac, and reported. His reply was brief. 'Report to your CSM and tell him you have passed out, and to give you a squad as soon as possible.' I had passed out in five days and it was another week before the others started to trickle back to report to Company HQ. Within days, I had my first squad of twenty-one recruits. They were a mixed bunch of good lads, all older than me. I was still 17, Harry Randall was my Superintendent Sergeant, with Trained Soldiers Scarfield and Meredith to get the 'shining' under way.

The squad did well, sailing through the various inspections. When we set a record score on the Empire test, Major Rice was so pleased, he treated the squad to a special tea — fish and chips, pineapple chunks and custard!

In 1941, German aircraft were running a shuttle service over the Depot, searching out the fighter stations at Kenley and Biggin Hill. Although they succeeded to an extent in their search, they didn't succeed in disrupting the smooth flow of training at the Depot. Collectively, we were all very macho and disdainful about the bombing and machine-gunning, but caught on your own without the reassuring presence of your mates was different. As squad instructors, we kept our recruits on a short rein at all times. One Thursday before tea, we had a warning of a raid. It was a heavy one and the sky was full of 'planes of both sides. Anti-aircraft fire was constant and occasional parachutes were seen billowing out. Dornier 'flying pencils' were down to rooftop height, their crews clearly visible.

My squad was considered sufficiently trained to help the defence of the Depot by occupying sandbagged strong points in allotments north of the officers' mess and adjacent to the main road. My orders were to speed to the guardroom with two recruits to draw ammunition from Scots Guards Police Sergeant George. As my two 'rooks' and I clipped along smartish through the allotments, dark shadows and deepening engine noise added wings to our speeding feet. The chatter of machinegun fire directly overhead made us dive for the doubtful shelter of runner bean rows where the arrival of three large bodies, plus a box of ammo did nothing for the beans' welfare and growing prospects. Back at the sandbagged redoubt, we watched as the 'planes passed over. Their targets were the airfields but some bombed the Depot and others diverted and attacked Croydon. Among the places hit was the Evening in Paris perfume factory, killing many young women workers.

After the raids, it was back to business with a vengeance to make up for the diversion. During one raid, I remember seeing the Depot QM, Coldstream Lieutenant Cooper (known to us as Duff) calmly cycling along, directing recruits to safety. He was a big man and his bulk and air of coolness inspired confidence in everyone.

The RAF 'erks' at nearby Kenley were having a fairly horrendous time in the day and night raids, and we were detailed to relieve them for much needed rest. The posts allocated to us were widely spaced. Doing my rounds one day, I cut across a corner of the landing ground and was caught with my nearest cover several hundred yards away. I dropped and tried to look like a piece of concrete. I'm sure my grey-green facial colour helped.

Lying there, I saw something I will always remember. Standing on the flat roof of a nearby building was a Scots Guards Sergeant in distinctive forage cap and shirtsleeve order, braces hanging over putteed trousers, manning a bren gun on a mounting - firing, unloading, loading and firing — giving a perfect demonstration of fire control, without any protection from the 'planes he was engaging.

A few nights after our return from Kenley, the Depot was bombed again and there was a direct hit on a building occupied by men of No 9 Company, Welsh Guards. 'Foggy' Farr's squad was wiped out. Seeing its members, or rather parts of their bodies blown into trees, roof trusses and adjacent fields aged him markedly.

After that, things eased up. Raids were directed more towards the capital, although dogfights continued and I saw one descending parachutist machinegunned from the air. Crews of shot-down German aircraft were often brought to the Depot and one Welsh Guards Sergeant who had been employed at Croydon Airport in civvy street recognised one captive as a member of a civilian German aircrew and a regular visitor to Croydon before the war.

Promoted to Lance Sergeant, I caught everything going — piquet sergeant, and so on, and the inevitable main gate guard.

Depot blackout regulations were strictly enforced and it was advisable to memorise the sentries' orders because you could barely see the board to read them. But I remember one night when, with the light from burning London, it was possible to read the orders without difficulty.

A final thought.

At Caterham, I discovered for myself what thousands of others have found — the genuine and enormously valuable bond of comradeship in the Guards. It crosses the regimental boundaries and you often find it takes you unawares when you might be expecting it least. My terrifying bogeyman, Mr Middleditch, was an absolute trump. After that early encounter, he helped me at every possible turn and he was among the first to celebrate my third stripe.

He welcomed me to the Depot Sergeants' Mess at the age of 18 years and one month."

———————oOo———————

Guards' Humour

The reasons are long lost in the mists of time, but it was always the approved method, in the small dark hours, of disturbing the sleep of anyone slumbering peacefully while you were on your way back from the latrines: 'Psst ... want to buy a battleship?'

———————oOo———————

What were you before you joined up?

Happy, Sergeant.

Blackpool to Bermuda

It was a journey which began at Caterham and ended four decades later and thousands of miles away in Bermuda.

Former Scots Guards Sergeant Ian Blackburn looks back on the journey now with considerable pride and not a little affection 'even though, at the start, it was quite frightening, and I had not the faintest notion of the rewards or punishments along the way or at the end of the road . . . '

"In truth, it really began before Caterham when I left home at Cleveleys, near Blackpool in March, 1956, to choose between being conscripted for National Service or signing-on for Regular Service. I passed the medical at the recruiting office in Preston, Lancashire. Then the recruiting officer asked many questions, including the type of service and the regiment in which I preferred to serve. Standing beside the officer was a ramrod-straight soldier wearing a blue uniform with a red sash, and a diced band round his cap.

Impressed, I said boldly that I would join his regiment. So it was done, and I signed-on for the Scots Guards, ordered to report for duty on April 6, 1956, at the Guards' Depot, Caterham, Surrey.

I walked up the long, steep hill from Caterham station to the Depot's main gate and, like all other recruits I noticed the tall, forbidding building with barred windows next door — the local asylum, which was to feature in much barracks humour about our true destination. Through the guardroom's open window, I announced my arrival for duty. Inside, a loud voice boomed 'RUNNER!', a cell door slammed followed by rapid footsteps, and a Guardsman shot through the door at express speed. He took off down the road past the square and towards the barrack blocks. The same voice bellowed, 'Follow that Guardsman, and do **not** overtake him!'

Terrified, I took up the pursuit, wondering what would happen to me if I lost sight of the Guardsman who was rapidly disappearing in the distance. I finally caught up with the Runner outside the barracks. Panting, I asked why I hadn't to overtake him. 'Because if you did, I'd be put on a charge!' My doubts about my Preston decision were already surfacing!

Depot training and discipline were both hard, but like so many others I learned to focus solely on what I was doing and life became, not just bearable but even enjoyable. I was innocently keen, however, and caught out at times. Like when our platoon sergeant asked if any of us could speak a foreign language. Two of us stepped forward. Then he asked if we would like to learn another language. Fool me, I stepped forward again. The sergeant's face bore a benign sneer: 'Your big chance. Now you can learn Polish — here are my boots!'

Never mind, I took my Regular status seriously and after Caterham, earned promotion. I did an eighteen-month stint as a platoon instructor at the Training Battalion at Pirbright. After that, I was transferred to the 2nd Battalion at Chelsea Barracks. During this period, my Battalion Trooped The Colour on Horse Guards' Parade, marking the official birthday of Her Majesty the Queen. It was a day I will always remember — all the pomp and ceremony, beautiful weather, the massed bands of the Guards regiments, and hundreds of Guardsmen in scarlet tunics. It was something to be a part of!

My service ended in March, 1959. So I took my Army Form B108 Certificate of Service and Testimonial ('He takes an interest in his work and is able to supervise others. He should do well in any future employment — he is both honest and sober') and followed my father into Government service. I spent nine interesting years in the Internal Revenue Department in Blackpool, Liverpool and London.

Then I took another of life's big decisions and came out here to join, firstly, the Bank of Bermuda and then The First Bank.

It's a world and forty years away now, but I can still hear that booming voice shouting for a runner as a terrified recruit stood shaking beside his battered suitcase at the old Caterham guardroom.

And I can laugh at myself."

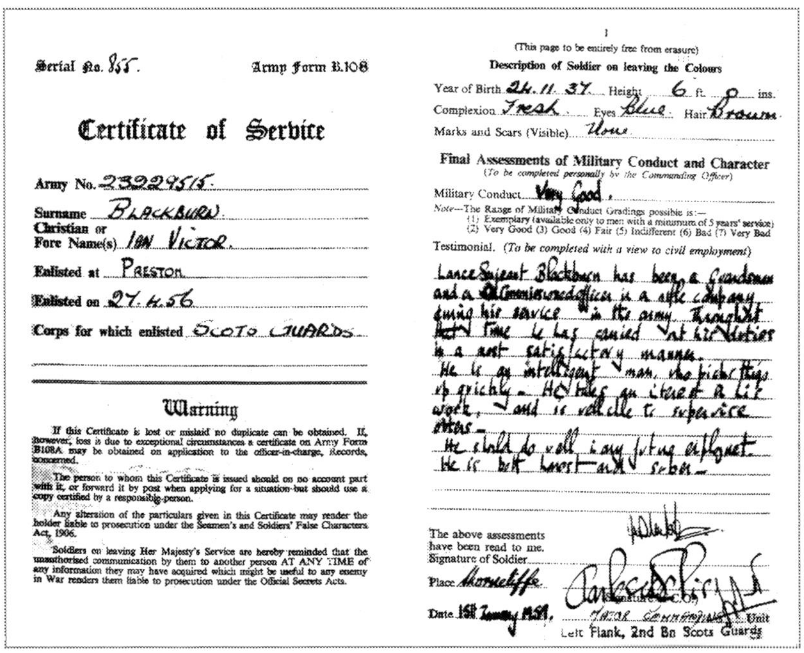

Serial No. 855. Army Form B.108

Certificate of Service

Army No. 23229515.

Surname BLACKBURN.

Christian or Fore Name(s) IAN VICTOR.

Enlisted at PRESTON

Enlisted on 27.4.56

Corps for which enlisted SCOTS GUARDS

Warning

If this Certificate is lost or mislaid no duplicate can be obtained. If, however, loss is due to exceptional circumstances a certificate on Army Form B108A may be obtained on application to the officer-in-charge, Records, concerned.

The person to whom this Certificate is issued should on no account part with it, or forward it by post when applying for a situation but should use a copy certified by a responsible person.

Any alteration of the particulars given in this Certificate may render the holder liable to prosecution under the Seamen's and Soldiers' False Characters Act, 1906.

Soldiers on leaving Her Majesty's Service are hereby reminded that the unauthorised communication by them to another person AT ANY TIME of any information they may have acquired which might be useful to any enemy in War renders them liable to prosecution under the Official Secrets Acts.

1

(This page to be entirely free from erasure)

Description of Soldier on leaving the Colours

Year of Birth 24.11.37. Height 6 ft. 0 ins.

Complexion Fresh. Eyes Blue. Hair Brown.

Marks and Scars (Visible) None.

Final Assessments of Military Conduct and Character

(To be completed personally by the Commanding Officer)

Military Conduct Very Good.

Note—The Range of Military Conduct Gradings possible is:—
(1) Exemplary (available only to men with a minimum of 5 years' service)
(2) Very Good (3) Good (4) Fair (5) Indifferent (6) Bad (7) Very Bad

Testimonial. (To be completed with a view to civil employment)

Lance Serjeant Blackburn has been a Guardsman and a Non Commissioned officer in a rifle company during his service in the army. Throughout that time he has carried out his duties in a most satisfactory manner.
He is an intelligent man, who picks things up quickly. He takes an interest in his work, and is well able to supervise others.
He should do well in any future employment.
He is both honest and sober.

The above assessments have been read to me.
Signature of Soldier

Place Moorcliffe

Date 1st January 1959.

Signature of C.O.

Major Commanding Unit

Left Flank, 2nd Bn Scots Guards

Sharper!

'Harry' Harrison had been 23688787 Grenadier Guardsman Harrison N R for more than five years when he first saw Caterham. He had joined in 1960 and missed doing his induction training there by about six months, going to the new Depot at Pirbright on the other side of Surrey instead. Then he was told to report to Caterham for the final nine months of his service.

It was a rude shock.

"I had served for nearly four years with 14 Company, the Grenadiers' training company, at Pirbright. Returning to the 1st Battalion, I was kitted out in full uniform and the officially-designated 'Change of Quarter Order'. This consisted of full webbing battle order, ammunition pouches to the front, two kit pouches at the back, full water bottle, rolled groundsheet, and on my back a large pack. I was carrying a fully-laden kitbag, topped by a steel helmet.

It all weighed as much, if not more than I did myself.

Luckily, I had battalion transport from Caterham station, up the dreaded and — for anyone on foot — the seemingly endless Church Hill, and past the St Lawrence's Hospital, the asylum which featured in so many caustic comments from irate drill pigs. I was dropped at the imposing and daunting main gate. Then my luck ran out.

Fully burdened as I was, I tottered as smartly as possible into the guardroom and 'tabbed in' to the Regimental Police Sergeant, who was sitting at his desk, writing. He didn't even look up and before I could announce myself, he barked: 'You're at Caterham now. Move yourself. Get out and get in again, SHARPER!' I got outside, put down my kitbag, took a deep breath, and returned — sharper this time. I came to an impressively shattering and abrupt halt in front of the Sergeant's desk. My large pack halted less abruptly and its continuing momentum almost knocked my forage cap off. Fortunately, my arrival announcement was accepted this time, although I was left in no doubt that I had arrived at an establishment which I had heard Caterham veterans describe as Sparta.

During my time at Caterham, the 1st Battalion was providing men for public duties in London, including the historic Bank of England Piquet. Before the Bank Piquet, we were inspected on the Depot square before being bussed up to London by Guards Coaches, a local company. Other public duties included the annual Trooping the Colour ceremony on Horse Guards' Parade.

One ceremonial duty I remember particularly was taking part in a Guard for the State Visit of King Hussein of Jordan, who had been to Sandhurst. It rained hard in London that day and ruined our red tunics.

Being a married man, I was provided with married quarters in the Barracks. My wife, two small children and I were quartered in No 2 Block, on the edge of a large green area. It was a cosy, two-bedroom cottage parlour style with an open fire (there was no central heating then). It was also furnished with everything from the usual table and chairs right down to the crockery and dish cloths. We didn't have to purchase a thing; it was all there.

Getting the battalion on parade at Caterham was quite a sight. The whole battalion would assemble on the side of the square, beside the drill sheds, standing about talking in small groups. Then would come the bellowed command from the Sarn't Major (RSM Tom Pugh), 'Parade!' Immediately, everyone would freeze in the stand-at-ease position. Not the slightest flicker of movement. The Sarn't Major again: 'Right. Let's have a good parade with no as-you-were's! Everyone waited, every muscle tight as a drum.

Then: 'Get on parade! As you were!!' — all in the same breath, before we had time to move. 'Get on parade. Get off again!' This time, we had managed to step off and half shoulder arms. 'You're all idle! Move yourselves, move yourselves, or I'll bag the lot of you!! Get on parade!!!'

On the word 'parade' the whole battalion stepped off as one man, shouldering arms as it marched. As if by magic, every man took up his allotted position on the hallowed square, transformed from a disorganised mass beside the drill sheds into an immaculate formation of Her Majesty's Grenadier Guards. Then would come the dreaded inspection when every man would hope that he 'rubbed', or passed scrutiny. You had to have prepared every detail properly because the inspecting officer usually the Adjutant, Captain JVEF O'Connell — would be followed through the ranks by the RSM, the Drill Sergeant (Drill Sgt H Knowles), the Sergeant-in-Waiting, the RQMS and the Corporal-in-Waiting, all with notebooks at the ready and eyes alert for a fingermark on brass, dust on a boot or — heaven forbid — a button undone.

When I was there, Caterham was no longer the Guards' Depot, but its standards were still as high and demanding as they had been at any time since 1877. And I'm sure they were the same when the last Guardsman marched out."

———oOo———

Guards' Humour

Some instructors would warn Caterham recruits: If you think it's tough here - wait till you see the assault course at Pirbright! The vultures wait in the trees all along its length!

———oOo———

Follow the Drum

Drummer John Tilbury, Grenadier Guards, joined as a boy. This is his story . . .

"Armies have marched across history to the staccato beat of the drum.

Over the centuries, relentless, urgent drumming sent nations' troops into attack in campaigns, remembered now by the battle honours on the colours of elite regiments. On the march, the tap of a side-drum marked the pace and kept units in-step and up to the required rate of progress. In peace, the drum had an important role in parades, from barrack square drills to ceremonials like guard-changing and Trooping the Colour — a part it still plays to-day.

Generations of soldiers have obeyed the call, Follow the Drum. And that's how I became a Grenadier.

I was 15 when I enlisted in 1937 so I joined on Boy's Service and was sent straight to the Corps of Drums in a battalion stationed at Chelsea.

There must have been something strange prompting my enlistment because, although I didn't know it at the time, a much older half-brother whom I'd never met — L/Sgt William Tilbury, Scots Guards, had been in charge of the Corps of Drums at the Depot from 1925 till 1931.

In true Guards tradition, my feet didn't touch the ground. My first afternoon at Chelsea Barracks found me learning to play the flute. That didn't excuse me all the formalities of drill and spit-and-polish. But in addition to these traditions, I was immersed in the process of becoming a musician — flautist, drummer and bugler.

I was stationed at Windsor at the outbreak of WW2 and was then posted to the Depot at Caterham where the Corps was in the charge of Sgt 'Pam' Greig and L/Sgt Bert Bridges, both of the Scots Guards. The Corps was made up of some 'elderly' drummers and some youngsters from all the five regiments.

The duties of the Drums were simple — a roster for barrack guards, bugle duty, providing music for recruit drills throughout the day, and for Commandant's Parade on Saturdays. Sgt Greig created march selections from many of the popular tunes of the day and squads did their 120-a-minute on the square to It's a Lovely Day Tomorrow, Run Rabbit, Run, Wish Me Luck, Roll Out the Barrel, Kiss Me Goodnight Sergeant Major, and We're Going to Hang Out the Washing on the Seigfried Line.

That last one was prophetic. In early February, 1945, many of us would be in the massive attack on the Seigfried Line, penetrating it and breaking out into Germany. At that time, our thoughts were not on laundry!

After Dunkirk, it was inevitable that war would reach us in Britain. It was a sunny Sunday afternoon when it reached the Depot. The Red Alert had been sounded in the barracks when bombers of the Luftwaffe flew low over us, on their way to attack Fighter Command's airfield at Kenley. As I leaped into a trench-like shelter on the cricket pitch, a young boy landed heavily an top of me, winding me and almost making me the Depot's first casualty. He was Derek Kirby, son of Coldstream Drill Sgt TW Kirby.

It is typical, I think, of the lasting friendships we formed in those difficult days that Derek and I corresponded until his recent death in Australia. Although his own service career was in the RAF, Derek always remained an ardent Coldstreamer.

The continued air raids changed life at the Depot. Red Alerts came thick and fast, bombs fell and there were many military and civilian fatalities. Bodies were placed in the Depot church. Pressures of war brought a huge expansion of the Depot with new guardrooms at Grove Road, The Fox and the Orderly Room. The Corps of Drums faced greatly increased duties as a result and bicycles were needed to get round the drill squares in time for parades. Training continuity became very difficult under the attentions of the Luftwaffe. Red Alerts seemed almost continuous at times — as many as fifteen in one day.

Invasion seemed imminent and we stood-to on a number of nights in trenches overlooking the area known as Happy Valley. With rifle, bayonet and five rounds of ammo, I don't know how effective a reception committee we would have made. Fortunately, it never came to that. Then the nightly bombing of London began and with it nights of living in shelters.

There were a number of direct hits on the Depot and the casualties mounted. The whole area suffered. It was in the line of the German bombers' run up to the capital.

At this time, the Depot had two out-stations. Old Coulsdon Drill Hall housed a company of recruits and a number of houses in Coulsdon Road served as billets. A mansion house in Whyteleafe housed another company which drilled on adjacent tennis courts. The Drums took part in all of them. Despite the increasing enemy action, the Depot social life went on.

Off duty, we were allowed out of barracks till 23.59hrs. Most evenings were spent in the local pubs — the Tally Ho, Caterham Arms, Surrey Hounds and the Old Fox.

Or we went for a bit of escapism at the cinema — the Davis Theatre in Croydon, the Regal and Astoria, Purley, or the Capital Cinema in Caterham Valley.

Drummer Tilbury J.E.

My highly regulated and ordered life at the Depot ended when my application to join a service unit was answered in September, 1941.

I was posted to the newly-formed 4th Guards Armoured Battalion on Salisbury Plain. A new chapter of service life was opening for me. I had no idea how many (or how few) chapters lay ahead.

But we all knew the final one would be the battle for the liberation of Europe."

Not the Best Start

The Scots Guards Sergeant in the Caterham Depot guardroom read the 17-year-old's joining instructions. Then he raised bulging eyes to the newly-arrived recruit, and half the town must have heard his anguished bellow: 'Robert Fraser McDonald — COLDSTREAM GUARDS? Why, you traitorous bastard ! ! '

Not the best start, wrote ex-Coldstream Guardsman (23206804) Robert McDonald from his home in Newton Abbot, Devon . .

"March 26, 1956 was a cold, grey day, which matched my mood when I arrived at the Depot, having travelled down overnight from Middlesbrough. The words of my father — a former regular in the Green Howards — were still ringing in my ears: 'You've joined the Guards? You won't last ten minutes!'

Then the encounter with the sergeant on the gate. (I'm sure his sash was wider than the sickly looking recruit before him). His 'welcoming' blast almost took my ears off and practically convinced me my father's prediction had been spot on.

Then he yelled: 'Get a move on and follow that man!'

My first reaction was Who? What? Where? Then I saw a uniformed figure setting off down the main barrack road. I made off behind him at my best speed. Over his shoulder, he told me I would be sorry if I overtook him. No chance! I had made up my mind on that score.

He left me at the reception centre. During the brief moment I had by myself, I thought: God, are they all like this? The Almighty answered my question with the arrival of a giant Coldstream full sergeant with more medal ribbons than Montgomery.

I went through my Caterham training without major incident, but the Depot and I didn't 'gel' and I was glad when I eventually left there. After that, Pirbright was like heaven. From then on, I enjoyed my soldiering and the Coldstream Guards.

Today, the mention of Caterham still haunts me. I have thought of re-visiting the Depot, but the thought of that giant sergeant of the guard has stopped me. Once is enough!"

In the 1800s

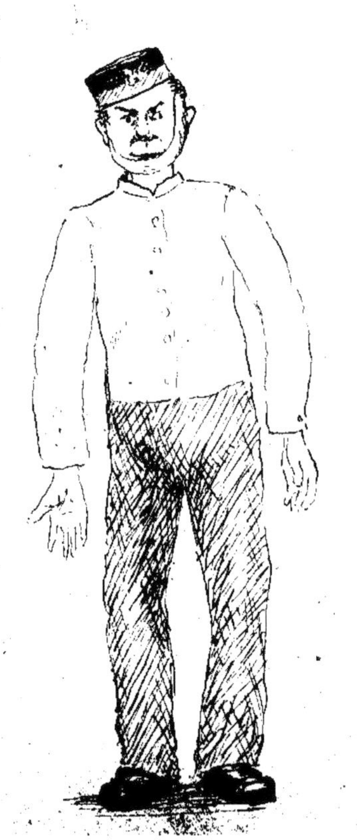

A Raw Recruit

A Tragic Ambition

This father-and-son photograph was taken at the Caterham Depot in 1903. In it, Colour Sergeant Sydney George Steer posed proudly with his son, Charles, who, even at the age of about eight, had ambitions to follow his father into the Scots Guards. Sydney Steer had joined the Guards from the Seaforth Highlanders in 1891, and fought in the Boer War. He had this photograph taken just before he left the regiment. He saw further service in the King's Own Yorkshire Light Infantry where he became an RSM. Later, he became a Court equerry to King George V.

His son fulfilled his promise and joined the Scots Guards as a boy in 1910. By the outbreak of WW1, he was a drummer. He was killed in action at Gheluvelt, according to the death certificate, on October 26, 1914.

He was shot, sounding a call and the bugle mouthpiece received his teeth marks as he died. The bugle was kept at the Caterham Depot till it transferred to Pirbright. For a time, it was lost. Now it is in the Guards' Museum at Wellington Barracks, London.

Drummer Boy Charles Steer with his father C/Sgt. Sydney Steer

In the 1800s

A senior officer's view of his second lieutenants.

He Was Always 'Blondie'

Hundreds of those who knew and served with him, and thousands who only heard of him as a regimental 'character' never used his real first name. Till the end of his service, he was Blondie Watson MM. His nickname came from his head of already fair hair, burned almost white by Egyptian sun in service just before WW2.

The MM came during that conflict in North Africa, on June 13, 1942, at a place near Tobruk called Maabus-Er-Rigel, when Blondie drove a truck full of wounded South African gunners to a forward ambulance centre while under heavy enemy small arms fire and tank shelling. His truck was hit repeatedly. And when he found enemy tanks had blocked his return, he fired his only weapon — a Very pistol — at the open turret of the nearest, which gave him time to escape and join a 25-pounder gun crew which had suffered casualties.

In his AB64 Part 1, he was 2695887 Guardsman Arthur Watson, Scots Guards, when he began his service at the Caterham Depot in 1938. That service nearly came to an abrupt end almost before it began, when a bizarre tragedy made the authorities believe he was dead. Twenty-one years later, he left the Guards at Chelsea Barracks, London, as ex-Colour Sergeant Arthur Watson MM, with war and peacetime service in Europe and the Middle East — a respected figure throughout his regiment and beyond — but all ranks still referred to him as Blondie Watson.

Towards the end of his service, he had gone back to Caterham to become an instructor — to put many recruit squads through the training disciplines he now regards as the foundations of his own life.

This is his recollection of how these foundations began to be laid.

"At the village of Whitehills on the Moray Firth, my life choices were going to be fishing or farming. So at the age of 16, I took the train to Keith in Aberdeenshire and the nearest recruiting office. The recruiting sergeant told me I was too big at 6ft 1ins for the Scots Greys, which was my first choice. I think he was on a bonus from the Scots Guards, so that was it.

After a medical and intelligence test, I was on a train that morning for London and Caterham with a rail warrant, 1s 6d (7 1/2p) ration allowance, and the King's Shilling.

Mid-day the following day, I was at the Guards Depot, Caterham. I was in a reception block where recruits were held until they were squadded in their respective regiments. In charge of us was an old soldier of the Irish Guards who we had to call Trained Soldier and stand to attention when we spoke to him.

All trained soldiers wore a Brigade Star on their upper right sleeves to identify them. We had medical inspections and very painful dental treatment (and equally 'painful' haircuts every week for a long time).

After two weeks, there were 12 Scots Guards recruits, so we were moved to Victoria Block to start a new squad, with Sergeant Jock Richmond as Superintendent Sergeant, Lance-Sergeant Smith as squad instructor, and TS Lawson in charge when we were not on parade.

It was about this time they discovered I was still short of the minimum age of seventeen-and-a-half. It seemed no one had checked my birth certificate, and I had kept quiet, of course. Then there was a tragedy.

A Guardsman with the same surname committed suicide. The authorities confused us and, for a time, I was officially dead. When the mistake was rectified, my real age was discovered. Thankfully, no action was taken. It was winter, but the fireplace at the far end of a huge barrack room was never used, although there was a large coal bunker beside it.

Issued with vast amounts of clothing, it was another couple of weeks before we had a squad of 18. Off-parade dress was canvas trousers and jacket with five regimental buttons which had to be fastened at all times. For parades, it was Service Dress and puttees — the worst item of clothing ever invented. It took us weeks to get them to the trained soldier's satisfaction.

There was no hot water for washing or shaving, but we had hot showers. Reveille was at 6am. We had to have washed, shaved with beds made up before breakfast at 7am. We had to have the barrack room cleaned and floor polished, wash house cleaned, and be outside on parade at 8.30am. We were issued with a buff belt, difficult to clean, and its studs had to be polished with a chewed matchstick and Bluebell. With little time to ourselves, the time flew past.

My squad passed the Adjutant's inspection after three weeks, gaining the right to be out of barracks between the Commandant's Parade on Saturdays till 10pm. Few of us took advantage for two reasons — there was too much cleaning and polishing to do, and we had no money.

Another six weeks on, another Adjutant's inspection and we were issued with the rest of our buff equipment, and capes with a valise star which was worn on a strap in the centre of the folded cape Then we moved on to an hour's religious instruction weekly, PT twice a week, and afternoons spent on weapon training and field craft. We were each issued with ten rounds of blank .303 ammunition in two clips. Something else to polish!

Everything continued to go well for our squad; our parade inspections were passed and our barracks inspections were usually OK. However, any mistakes in the barrack room were punished immediately, usually by having to run six times round the training track in the evenings, equal to one-and-a-half miles. Parades finished at 4pm, tea followed at 4.30pm, and then it was straight back to the barrack room for shining parade when we cleaned and polished our equipment and were force-fed regimental history, battle honours, names of commanding officers, and recognition of bugle calls. Shining parade ended officially at 7pm, but frequently went on till lights-out at 10pm.

We were well versed in sentry and picquet duties by this time, but after the 12-week inspection by our Company Commander, we took our turn on guard at the barrack gate. I was in the first guard from my squad — quite an honour.

At Caterham, our pay was seven shillings (35p) a week. From this, three shillings was kept in credit, against any possible damage to, or loss of kit. The remainder was available as leave money.

We had three substantial meals a day, except Sundays when the evening meal consisted of two slices of bread with butter and jam, plus a slice of cake. And with all the drill, PT (and many times round the punishment track), we were not only slim, but very fit.

The Commandant's inspection after 16 weeks was passed (to our astonishment) without difficulty, and there we were — Guardsmen, after a total of 18 weeks of sweat and tears, if not bloodshed. It was all over. We left the Guards Depot about a week later, hoping we would never see the place again.

Then we joined the 2nd Battalion to sail for Egypt."

Colour Sergeant Arthur Watson MM when a Superintendent Sergeant at Caterham

In the 1800s

Corporal - ' Why, there's enough grease on yer collar to make pies for the 'ole mess.'

Shifting Attitudes

In terms of strict chronology, the Coldstream are the second-oldest regiment in the Guards Division. But their defiant regimental motto is *Nulli Secundus* — Second to None. And that's how 23342507 ex-Lance Sergeant Keith Murrin of Exmouth remembers his two-year National Service, which began at the Caterham Depot in October, 1956 ...

"To this day, I still have my rail ticket to Caterham. And I clearly remember that dull, October afternoon when I first saw the awesome sight of the Depot main gates. I had been called up for National Service, and I checked in at the guardroom, not knowing what to expect.

Very soon, I began to wish I had taken my father's advice, applied for a deferment, and given the Coldstream Guards a big side-step! But I was in, and that was that.

The early days, with all the strange and seemingly unfair restrictions and disciplines, were off-putting to say the least. Having to stand to attention and ask the Trained Soldier for permission to go to the toilet, was just one example. But although we didn't realise it at the time, our views and attitudes were shifting gradually under the grind of the training programme. It happened very slowly, but by the end of eight weeks we felt a sense of pride that we could take anything they could throw at us.

Shining parades were supposed to be from 7pm to 8pm, but they seemed to go on till 'hands and feet' inspection at lights-out around 10.30pm. Visit the NAAFI? That was unheard of!

One vivid memory is of a Commandant's Parade on a Saturday morning about January, '57, when all the Depot recruits were on the square. There were at least a thousand of us, and we had been lined up in strict seniority to await inspection, our squad under the wary and watchful eye of Corporal Heywood. We were about ten or eleven weeks squadded and feeling quite proud of ourselves.

It was damp and cold, both of which we were feeling when the officers came on, to march up and down and receive the salutes of the Grenadier RSM and his two drill sergeants in turn. Then the RSM proceeded to call the parade to attention six or seven times, only to shout 'as you were!' each time. It was all very frustrating. So much so that, although talking in the ranks was punishable with all kinds of very unpleasant penalties, a voice somewhere was heard: 'Sergeant-Majors', it said, 'take them away they have too much sleep and too much to eat. Bring them back when they are ready!' Fortunately for the recruit responsible, he was never identified.

After being 'rifted' round the square for twenty or thirty minutes, we were formed up behind the band and marched out of the Depot and along the main road, before returning.

We survived the Depot and I went on to the Training Battalion, and then to join the Coldstream's 1st Battalion. I can say — hand on heart — that I thoroughly enjoyed my two years in the Guards.

I have re-visited the Depot on two occasions,

and if a Caterham reunion is ever held, I know a lot of old-timers like me would turn up in droves to re-kindle memories. If any of the old buildings are to be demolished under the future development plans, quite a few would buy a brick — proceeds to go to charity."

——————oOo——————

Guards' Humour

A former Scots Guards Warrant Officer took up a retirement job as Official Starter on the first tee of one of Scotland's major links courses. From his box, his job each day was to book golfers in for play and then call them forward to the tee when the time came for them to drive off. As most players were either Scots, English or American, there were few problems of communication. Came the day, however, when a French enthusiast - a Monsieur Fouquier - turned up and requested a starting time for his round. There followed a slightly laborious process in which M. Fouquier spelled out his name, using the French pronunciation of each letter, and finishing with the phonetic pronunciation — FOOK-EE-AY.
Throughout this recital the Starter glowered at his ledger as he entered the details in his schedule of play. Then, fixing the Frenchman with a withering glare, he announced: 'Right, sir. Ye'll play off at 10.30am, sharp — AND YE'LL ANSWER TO THE NAME OF FERGUSON!

——————oOo——————

Coldstream Guards

Father Knows Best?

When 17-year-old Stuart Fisher decided to sign on in the Guards, his ex-Welsh Guardsman father said: 'Join any of the other regiments — but not the Welsh Guards. Stuart knew his father had tried to join the Coldstream in 1922, but was two inches short of the six-footers they were looking for. He had been passed on to the Welsh Guards, where the height requirement was not so demanding.

"Having sworn my attestation as a regular at St Albans recruiting office in early September, 1945, I was given a single rail ticket to Caterham and caught a bus to Caterham-on-the-Hill. At the Depot guardroom I was checked over and spent a few days in the Civilian Reception Centre. Then I was taken to Monck Block where six-foot-six CSM Frank Vonk grabbed me by my civilian jacket and pulled me into the Company Commander's office. I was taken aback. (The CSM was undoubtedly unaware that I had spent seven part-time years in the Church Lads' Brigade and the Army Cadet Force, and was used to foot drill).

My being English, Major Makins, the CO, asked why I wanted to join the Welsh Guards. 'Because,' I said — boldly ignoring my father's advice — 'my father served in the Regiment, Sir!'

Then I became 2741297, a number which will live with me forever. Out of 20 recruits in our squad, only three were regulars and we were all English. Des Boak from Redcar had also been in the Army Cadets. He stayed in and became a Regimental Sergeant-Major. The other regular was Dick Harry with whom I later lost touch. We slept in a set of PT kit and used a towel as a pillow slip. It wasn't until 1946 when we joined the 2nd Battalion in Germany that we had sheets. Pyjamas came in 1949 when we were training for Malaya (where we didn't go!).

At Caterham, we must have been innocents. When CQMS Nichols advised us, during a lecture, that we were not to get any girls into trouble on the Common, we didn't know immediately what he meant!

The Depot characters I remember were CSM Vonk, a great chap who had been taken prisoner when France fell; weapon training Sergeant Danny Bartlett; Sergeant 'Greased Lightning' Bell (everything had to be done like that!); and Sergeant Webb, a great guy who was a POW, but escaped to the Russian lines and earned the DCM. Our Superintendent Sergeant was Dan Daly, who chased us mercilessly on the square (I learned to do the Guards' about-turn on the march, rather than the one I learned in the Cadet Force). Sergeant Griffiths was our PT instructor. He told us he was a hangman in civvy street! I believe it, because he used to leave us hanging on the wall bars till I'm sure we stretched another couple of inches.

On one rare occasion, I was awarded a credit on parade and was allowed a week-end pass. It was my only outing from the barracks in four months. When I returned on Sunday night, the Grenadier Sergeant of the Guard told me: 'You've come in the wrong gate, lad'. The penny dropped slowly because it was my first brush with the ancient Depot joke about the lunatic asylum being next door. Saturday mornings, we would go on a 'swank' parade, when the regimental band played and we weren't chased. Sundays, we went on church parade, and it was then I first heard that lovely Welsh tune, David of the White Rock. It has been one of my favourites, ever since.

The Welsh Guards' battle honours had been painted in regimental colours on our barrack room walls. These had to be wiped over with an oily cloth to make them shine for room inspection. The beds were in two sections, the bottom one sliding under the top one when we weren't in them. The only reason we were allowed to go to bed, we were told, was to press our battledress by lying on it.

For shining parade, we spread our groundsheets over the beds to blanco our belts, gaiters, and rifle slings. Our brasses had to glow and our cap badge, the Leek, was rubbed on bath brick to smooth it and improve the shine. We had short lengths of wood on which to polish our chinstraps and gaiter straps. During shining parade, we were taught our regimental history, which began in 1915 when the regiment was formed. It could have been worse — the Grenadier, Coldstream and Scots Guards were formed in the seventeenth century and had been in action all over the world! The Irish Guards had been going since WW1.

The jargon of the Depot took hold quickly. Why sausages were called 'swanks', nobody knew, but they always were. I remember one great chap who was known as 'Swank-lips' Jones. It was always 'tea and wads', and something you'd overlooked was compared to 'Scotch mist' - as in: 'What d'you think these tapes are, lad, Scotch mist?' In the Welsh Guards, anyone named Evans, Griffiths, Williams or Davies had to be known by surname and the last two digits of their regimental numbers to avoid inevitable confusion.

Eventually, after sixteen weeks squadded at the Depot, we had a lecture by our CO in which he told us never to let our Caterham standards drop — and we were off to the Training Battalion, which at that time was stationed at Sandown Park Racecourse in Surrey. The horses had either been moved out, or eaten. We were allocated one stable to six men.

But that's another story . . . "

Thirty-eight Years a Guardsman

To the thousands of recruits who arrived at Caterham from 1938 till the Depot moved to Pirbright in the 'sixties, Sergeant RF George, BEM, Scots Guards, was the dramatic, physical embodiment of all the attributes they themselves would be expected to strive for as Guardsmen.

Their first sight of Sergeant George was certain to impress, if not terrify. This towering, immaculate figure of gleaming boots and razor-sharp creases had a granite face, the profile of which had been fashioned, apparently, by his Maker to complement and continue the downward plunge of the 'cheesecutter' or SD cap traditionally worn in the Guards.

The fact that he was the Depot Police Sergeant did nothing to relieve the apprehension recruits felt about Sergeant George. Many, however, learned to their advantage that although he was a strict disciplinarian, he always tempered justice with mercy. Chancers, on the other hand, got no quarter.

When Richard Frank George became Police Sergeant at the Depot in 1938, he had already been in the regiment for longer than the incoming recruits had been alive. He died in 1976. But in 1959, with retirement looming, he had put down some memories of his 38-year service for the Scots Guards Magazine.

He recalled the day in 1922 when he tried to sign on. But the King's Shilling was not to be had that easily. At least two character references were required, one from his school. He had to await the arrival of these testimonials and spent the time carrying chairs from Whitehall to the Mall to seat guests to the society wedding of the year - that of Princess Mary to Lord Lascelles.

On acceptance, he reported to the guardroom at Caterham, bright-eyed and full of youthful enthusiasm. The duty sergeant took his papers, looked him up and down with a jaundiced eye, and treated George to a piece of black and politically-incorrect humour every recruit was subjected to from the day the Caterham depot opened, till the day it closed.

"Boy," he said, "you've come in the wrong gate." St Lawrence's Hospital for the mentally disturbed was next door.

George wrote: 'Many times in the next few months, I felt I might have been better off if I had taken his advice!' His first meal was an offering, edible only with the pangs of extreme hunger. His first ablutions were with cold water in the old tin bath house which had no glass in the windows. His first haircut was a scalping.

'The trained soldier got a move on us straight away. All instructions were shouted: 'Get your buttons sewn on your service dress jackets, your greatcoats, and your canvas jackets. You are in Sergeant Alec Ross's squad and I am your Trained Soldier, so watch out. Your company commander is Captain Eric Mackenzie DSO, and don't forget it!'

For George, a bad start got even worse. A bawled order, just before lights-out: Stand by your beds! And although he leaped to it, he was not quick enough to satisfy Sergeant-in-waiting Freddie Archer, one of the most terrifying of the Depot's NCOs.

'What's your name?'

'George, Sergeant.'

'George, what?' I then repeated my name. Archer glared.

'Have you holes in your socks?'

'No, Sergeant.'

'Then how the hell d'you get your feet in

them?' I thought I was expected to join in the joke, and laughed. That did it. I was 'in the report' for the first, and by no means the last time. The next three weeks I remembered for blistered feet, aching legs, the horrendous complexity of getting puttees right, and the manner in which we did the unending drill. And until we passed out on foot drill, we went everywhere around barracks with our arms stiff at our sides.

After twelve weeks, George had permission to 'walk out', something he, like all recruits, lusted after throughout the initial incarceration in the Depot. But walking out involved barrack gate inspection, a hurdle to be cleared before recruits were revealed to the public gaze. Three times, George presented himself at the gate — and three times he was turned back. 'I resigned myself to the thought that it probably wasn't worth going out anyway.'

It was to be another twelve weeks before he saw the outside world again. That was when the Colonel of the Regiment, Lord Methuen, decided that George's squad along with other fledgling guardsmen who had completed their basic training, were fit to join a battalion.

With the 2nd Battalion at Aldershot, George learned the rituals and complexities of the ceremonial 'Buck House' guard duties he was to do in 1923 when his unit moved to London. At that time, the sentry boxes were outside the Palace railings, so that sentries had to learn to cope with the mischievous and even provocative behaviour of some who came to watch. When things began to get too difficult, the correct course was to snap to attention, shoulder arms, and march up and down the sentry's 'beat' till the offending spectators gave up or an NCO arrived to post the sentry's relief.

Three years later, in 1926, the 2nd Battalion was moved to Warley Barracks, Brentwood, and George gained his lance-corporal stripes. 'That same year found us working on the London Docks throughout the General Strike. Then the Battalion returned to Aldershot, and I went back to Caterham in 1927 as an instructor. That posting didn't last long, and I was ordered back to the Battalion on its embarkation for China later that year. Back from China in 1929, I joined the Depot at Canterbury as a PT instructor.' Between 1931 and 1938, George was with the 2nd Battalion at Windsor. Then he returned to Caterham where he witnessed the convulsions which overtook the Depot on the outbreak of WW2.

'Looking back over these eventful years at Caterham in the early years of the war, I remember the opening of the new camp in 1940, when sixteen companies were situated in and around the Depot, and all the empty houses on the road to Purley were occupied by squads. Gone was the Service Dress and the hated puttees. Forming fours, piling of arms and forming squad became things of the past, and Battledress — the practical, active service uniform which was to be worn by the British Army throughout the war — was introduced.'

Intensive training throughout the Army was urgently necessary, and the quality of the Guards' basic regime became a model for all to follow. Warrant Officers and NCOs from almost every regiment in the Army attended drill courses at Caterham. The thousands of wartime conscript recruits who passed through Caterham knew Sergeant George and recognised that his reputation for strictness with fairness was justified.

'With war's end came yet another upheaval at the Depot. Most of our old officers and senior mess members returned to civilian life, and 1947 saw the arrival of 600 National Service recruits, the first of many still to come.

'With the inevitable drop in recruiting after the war, the numbers of recruits joining at Caterham fell sharply. There was no longer any need for the hutted camp, and it was slowly demolished. I find it hard to believe that where we now have green fields, there were once three great parade grounds where thousands of recruits sweated it out under instruction.'

Sergeant George served at Caterham under nineteen RSMs, and he retained memories of every one of them. All of them remembered and valued his support. More than thirty years after he first entered the Depot Sergeants' Mess, he wrote: 'My service in the regiment seems to have passed all too quickly. I hear that CSM Bertie Lemon's grandson is now joining the regiment. It makes me realise just how quickly the past thirty-eight years have gone.'

In the 1800s

"The. Pride of the Depot." May 1890.

'Titches' and Stitches

The tailor's shop at Caterham determined the basis of what degree of sartorial correctness recruits achieved at the Depot. Each influx of new men had to be turned out in uniforms, suitably altered and adjusted to withstand the smouldering, serial examination of squad instructors, CSMs, RSMs, and officers. This daunting task is recalled by 2741601 ex-Welsh Guardsman Raymond Jones who now lives in Whittle-le-Woods, Nr Chorley, Lancashire.

"The master-tailor was 'Titch' Risebrook who was still around and sporting WW1 ribbons in the 1950s. His staff consisted of 'Titch' Miles who had just been promoted after thirty years as a Lance Corporal, 'Titch' Evans, a Lance Corporal in the Coldstream, and a Welsh Guards Trained Soldier who had served a measly nineteen years. The first three had been dubbed 'Titch' because they were all below the average depot height, but there was a lot of service as well as expertise in the shop!

'Titch' Risebrook measured the Corps of Drums (of which I was a member) for tunics and these splendid items duly arrived and were fitted. The Adjutant (I think it was Captain Jack Scott) was required to inspect the result. He seemed satisfied — till he got to me. Then he confirmed my own feeling that my tunic didn't fit all that well. And he made sure his own displeasure was not missed, stressing that more care must be taken with measurements.

The ever resourceful Risebrook whipped out his tape measure and quickly checked me over. 'I have it, Sir,' he reported. 'This man is deformed!' He had discovered something even I didn't then know — I have a dropped shoulder.

My old boss in the Corps was Grenadier Sergeant 'Gerry' Fisher, an extremely talented musician — pianist, flautist, and composer. He was also the Depot bookie, whose motto and advertising slogan was 'Fisher for Fairness'. He was a lovely man. I had the privilege of serving with him when I was a drummer, and in a further tour of duty a few years later when I was an NCO.

He composed a march entitled 'Les Deux Colonels', which reflected the fact that both the Quartermaster and Commandant at that time were Lt. Colonels. I have often wondered if this was a subtle piece of 'flannelling' to encourage a Nelson's eye view of his slightly irregular sporting activities!"

In the 1800s . . .

On mess fatigues

The Watchers

There are those who will tell you that many of the thousands of men who went through the rigours of the intensive induction training at the old Caterham Depot left an almost tangible presence which can still be felt around the empty barracks blocks, the old guardroom, the windswept square, the drill sheds, and the silent roads.

Some who went back to make farewell visits to the abandoned Depot have claimed to have heard the faint, indistinct voices of Guardsmen who fought in half-forgotten wars and others who soldiered through campaigns of more recent times. Individual experiences of this kind can be attributed, perhaps, to tricks of the wind in deserted corners, or echoes distorted by empty distances and amplified by the intensity of the listener's own memories of the Depot.

There is, however, one recorded and unexplained incident which was experienced by a group of men at the same time. Former Grenadier Malcolm Found, whose father and grandfather served in the regiment, tells the strange story . . .

"It began early in 1996. The Honourable Artillery Company was about to be presented with new Colours by their Captain General, Her Majesty the Queen, and there was an urgent need to ensure that arms and foot-drill on the occasion was of the highest order. The then Adjutant Captain RAC Duncan and Sergeant Major Alan Prentice felt there was no better place to conduct practice drill parades and rehearsals than the square at the, by then, abandoned Caterham. They sought and received permission.

The desired results were gradually achieved. Thus, one week in May, the Honourable Artillery Company's Corps of Drums arrived at the Depot to take the Guard of Honour through a last rehearsal. The final phase of this required the Guard of Honour to march down the main Depot road behind the Corps of Drums. With a repertoire of more than thirty marches to choose from, Drum Major Keith Hingle — a former Coldstream L/Sergeant — had selected The Guards Return as appropriate to the day and moment. It is a tradition that this march is played as a Guards regiment marches back into barracks.

As the parade marched through the deserted Depot towards the guardroom, almost every drummer felt strongly the presence of invisible watchers. The feeling continued till the end of the parade. Nobody mentioned the experience then, each man believing he alone had felt the 'presence'. The fact that almost everyone in the parade had had the same powerful sense of being watched emerged in the bus on the way back to London. Drum Major Hingle's feeling was typical: 'During the last few hundred yards of the march towards the main gate, the hair on the back of my neck stood up. The feeling of an unseen presence was unmistakable.'

None of those at the Depot that day had experienced anything like it before, and according to reports, none have since. But to a man, they are adamant that the purpose of the parade, the music and the place had combined somehow to entice the shades of generations of former Guardsmen from the old Depot's bricks and mortar.

—————oOo—————

Guards' Humour

With a few painful Caterham weeks behind him, one recruit drew up concise advice for his squad, and all those who would follow:

1 - The Trained Soldier is always right.

2 - When the Trained Soldier is wrong, re-read first point.

—————oOo—————

Don't Look Now, But . . .

Rooted in the solid, hard-nosed daily practicalities of no-nonsense Guards' life, Caterham Barracks seems an unlikely setting for a brush with the occult. But this recollection from ex-Welsh Guardsman Gordon Coleman, now living in Valentine, New South Wales, will raise short back hairs on the necks of recruits who did sentry duty at the Depot's rear gate when winter winds blew and flickering gaslight fought fitfully with the surrounding darkness ...

All began routinely enough, recalls 2741389 Coleman, G:

"I did my Caterham 'sentence' from November 1945 to March 1946 — with no remission for good conduct. Our squad was lucky to have two good blokes in charge, Sergeant Peregrine and Corporal Gillam. Of course, when we were suspected of playing up, or falling down on the progress we were expected to make, we got our share of chasing — 'rifted' around the drill square till the traditional jibes about having 'come in the wrong gate' seemed all too true.

And the fact that the asylum was close by the Depot back gate did nothing to lighten the atmosphere for the lone sentry, posted in the lane outside the Depot, when the two heavy wooden doors closed behind him. This lane was a very creepy area, a long way from the guardroom. The sentry box stood beside an old gas lamp with a flickering mantle. On windy nights, the light would go down so low that all light disappeared; then it would flare back to life for a few brief moments.

The wind also moaned eerily through tree branches which scraped on top of the sentry box. When the sentry patrolled his 'beat', a resident echo sounded as if someone (or something) was creeping up behind. Needless to say, the NCOs who posted green sentries did nothing to dispel the Hammer House of Horror atmosphere which descended when possible escapes by the unfortunates next door had been discussed, the doors creaked shut and were padlocked, and the footsteps of the departing NCO and the old sentry died away.

Now, to lighten the duty and help pass the time, some recruits on sentry-go had befriended a cat which frequented the area after dark; it would arrive and perform feline weavings through the at-ease sentry's legs, and sometimes offered itself for highly-unregimental cosseting in the sentry's arms. That is, until Recruit Morgan did his first sentry duty (there were three Morgans in the squad, and I won't give this one's number).

This Morgan was standing by the sentry box and wishing he was back in the comforting light of the guardroom when a black object hurtled out of the darkness and landed on his chest. Morgan let out a terrified yell and fell back into the sentry box. This commotion, and the whistle on a lanyard all sentries carried, brought the duty NCO down to the rear gate at the double.

There, he found a terrified cat — and an incoherent, babbling old man of nineteen. Concludes ex-Guardsman Coleman:

This is a true story. Only the name of the cat and Morgan's number have been withheld to protect the innocent."

———————oOo———————

Guards' Humour

A Coldstream Sergeant adapted Kipling's 'If' verses: If you can keep your head, while all around are losing theirs - maybe you don't understand the situation!

———————oOo———————

Stand By Your Bed — Wherever it is!

The administration of justice in the Guards is governed by a strict code, just as it is in the Army's line regiments. It is formal and official, and any attempt to take unofficial action is likely to be met with severe punishment. But as every Guardsman knows, the justice of the ranks is still carried out, although it is never 'seen to be done', like its official counterpart. It often takes very strange and unorthodox forms. This example was witnessed by a Grenadier boy recruit, Roy Cox, who entered the Depot just after WW2 had ended.

"As usual, there was a huge amount of rivalry among the Boys' Company squads, each doing their damnedest to be the best, the smartest — and to match up to their instructor's progress targets. And woe betide anyone who let the squad down!

To the disgust of everyone, there was one lad who found it difficult to come to terms with the discipline, and was more than once the cause of 'back squadding' — being made to repeat a week's training. Anyone who was at Caterham will tell you how serious that was. The situation came to a head one evening, when word was sent around that we were all to gather in the Irish Guards squad hut for a barrack room court martial.

We duly gathered as requested in the late evening before 'lights out' when there were no Instructors or Trained Soldiers around (I wonder why!). Two boys were appointed, one for the prosecution and the other defending. The accused was duly found guilty. He was sentenced to Run the Gauntlet, which meant him running the length of the hut, through the rest of us, armed with slippers. The punishment wasn't very successful. Shortly afterwards, he let the squad down again.

Plan Two was agreed.

A few nights later, after 'lights out' and getting on for midnight, a check was made to ensure that the offender was sound asleep. Then the rest of the squad transported all his equipment, his locker and his bed (with him still asleep in it) to the centre of the nearby sports ground and left him there. It was only when reveille sounded at 6am that he realised his predicament. To give the lad his due, he managed to get everything back in the hut before breakfast, but only just.

Plan Two seemed to do the trick. Although he never became the best of soldiers, he managed not to let the squad down again.

I completed a further tour at Caterham, five years as a side drummer in the Depot Corps of Drums. I became the Adjutant's side drummer, which meant I had to attend all his parades and those of the Commanding Officer, beating out the time for arms or foot drill, with the aid of a metronome and under the direction of the Regimental Sergeant Major.

During the last years at Caterham, I met my future wife, June, who was living at South Merstham.. To do my courting, I bought a bike and, together with my Drummer mate, George Lippiatt, cycled most evenings to Merstham where he, too, had met his future wife, Ida.

Bikes were really a necessity for the duty drummer at the Depot. He needed one to get around to sound the bugle calls on time (or near enough) which the Depot was dependent on for the timing of the various duties and parades. George and I set up a sort of bike repair shop in our block. We had no need to buy spares — they were readily obtainable from the council dump at the back of the barracks.

After we married, June and I were quartered in a two-bedroom flat at the Grove flats, which were then part of the Depot. We had to furnish it, and although there was a so-called water heater, we had to supplement this by boiling water on the gas stove to have a hot bath. Like two other families in the Grove flats, we kept bantams in our small kitchen garden. Several times, they escaped and June was usually the one who had to get them back.

All this is nearly a lifetime away, but it was an experience I would not have missed."

Grenadier Guards

Operation Westminster

The Guards occupied the House of Commons on September 11, 1950. Hundreds of Guardsmen, drawn from Caterham, London, Pirbright and Windsor took over the Mother of Parliaments.

It was a silent operation. In fact, they did it in slippers!

One of the many historic buildings badly damaged in the London blitz of WW2, the Commons was substantially re-built when peace was restored. In the process, a number of improvements were made, notably air-conditioning and enhanced acoustics. To test both, the chamber had to be occupied as if Parliament was sitting.

So the Guardsmen — supplied with slippers to protect the new carpets — filled the Government and Opposition benches as well as the public galleries. Donald Thompson, a 21-year-old Coldstream Guardsman, stood in for Prime Minister Clement Attlee. A 19-year-old Irish Guardsman, Ronald Chicken, from Liverpool, sat in Winston Churchill's seat.

Lord Morrison welcomed the Guardsmen as 'new Members'. Grinning, he said he was delighted that the country had come to its senses at last and had elected a sensible, intelligent Parliament.

MPs for the day, the Guardsmen moved through the lobbies and round the galleries, taking part in a series of experiments to check the new sound system, built into the benches, and the clarity of the acoustics.

The House rose in the afternoon to allow the temporary Members to return to their 'constituencies' in Caterham, Pirbright, Windsor and London.

SECTION C.D

A Buck House Shock

Edward H Bradbury did his Caterham training in 1952, following his father and several cousins into the Grenadiers. After Pirbright, he was posted to London to do public duties from Wellington Barracks.

He will always remember taking part in the Coronation in June, 1953. Then Queen Mary died, and he received an order which almost got him into a lot of trouble. Now living in Burntwood, Staffordshire, he recalls —

"I was detailed off to collect the pictures of Queen Mary from the guardrooms at Buckingham Palace and St James's Palace and deliver them safely to Wellington Barracks. When I marched through the gates of Buckingham Palace and approached the guardroom, I encountered three men talking in a group — a Guards officer and sergeant, and a man I couldn't put a name to, who was wearing civilian clothes.

I marched up to them, came to attention, and saluted. They went on talking.

I wanted to carry out my orders, so after a time I spoke in my very rough and raw Black Country accent. All three turned and looked at me, and if looks could kill, the sergeant's certainly would have. He gritted, 'One doesn't speak till one is spoken to.'

Now I'm for it, I thought. But when the officer and the civilian smiled at me, then laughed as they walked away, his expression softened. 'All right, lad, relax. You've got away with it — you made the Duke of Edinburgh laugh.' I quickly retrieved the picture from the guardroom and left, even more quickly.

Once a Grenadier, always a Grenadier. They were hard times, but happy times, too ..."

In the 1800s . . .

A Scots Guards recruit AF Emery included a self-portrait on the right of his tit-bits collage of Depot characters.

Guardsmen in the Garden

A firing squad drawn from the Brigade of Guards carried out public executions in London in September, 1946.

The involvement of Guardsmen in carrying out the death warrants was due directly to a personal request expressed by Queen Victoria in the mid-1800s.

The event which put members of the Brigade in the execution squad took place one night when the Queen and Prince Albert were attending a performance of the Beethoven opera, Fidelio, at the Royal Opera House, Covent Garden.

At one point in the opera when soldiers march on stage, the Queen expressed surprise and not a little displeasure that their physique and bearing was not to her liking and that they were not marching in time to the music. She sent for the director of the Royal Opera House and told him stingingly: 'My Guardsmen can do much better than that!' Then she decreed that men of the Scots Guards should fill the non-singing parts in future operatic productions.

Her Majesty's 'request' was later broadened to include the other four regiments in the Brigade.

So it was that in 1946, the firing squad which executes the ill-fated hero, Cavaradossi, in the final act of Puccini's Tosca, were Guardsmen serving in London and Caterham at the time. In eight performances of the opera that September they fired 'the fatal volley'.

The world's leading operatic tenor of his day, Benjamino Gigli appeared at Covent Garden with the San Carlo Company at the end of WW2. So did many Caterham and London-based Guardsmen, who filled non-singing parts in all the featured operas.

Singing the part of Cavaradossi was Beniamino Gigli, Caruso's successor as the world's leading operatic tenor, who headed the San Carlo company from Naples in a London season. Over a drink offstage, he thanked the Guardsmen for taking part, and gave a signed photograph to the man filling the role of the officer commanding the firing squad.

Guardsmen took part in at least ten operas with the San Carlo company that year in a season which ran to fifty performances. Among them were Carmen, where they were toreadors, and La Boheme, The Barber of Seville, Rigoletto, Cavalleria Rusticana and Pagliacci where they took part in crowd scenes.

The tradition linking the Guards with Covent Garden was a popular one and there was never any problem finding volunteers for the duty over many years — not least because it was a useful way of supplementing Army pay.

Sadly, the tradition ended in 1978 when the actors' union, Equity, insisted on the Opera House employing actors to fill the non-singing roles.

——————oOo——————

Guards' Humour

Anonymous advice to Caterham recruits: When the Drill Pig stops in front of you with eyes bulging, face purple and with froth around his lips, pay attention. These are some of the little signs that tell the keen observer that all is not well.

——————oOo——————

Shocking, Sergeant Major!

The scandalous affair of the dropped trousers swept through the Depot one day in the 1930s and was celebrated by recruits long afterwards in song and story.

It was said to have raised Regimental Sergeant Major 'Tibby' Brittain to a state of near apoplexy by reducing one of his parades to a state of chaotic hilarity.

It began innocently enough one day, when the Depot's Trained Soldiers were put on orders to attend medical parade in the gym. It was supposed to be one of the regular and routine inspections, but what followed was far from routine.

RSM Brittain drew the Trained Soldiers into an orderly rank along one of the gym walls, before the arrival of the medical officer and his team. The MO and his entourage duly arrived, accompanied by the duty Sergeants-in-Waiting who, as usual, eyed Brittain with some trepidation.

As the MO began his inspection and moved along the rank, the only sound was that of braces hitting the floor, punctuated occasionally by a short, subdued conversation as someone had his name taken and was ordered to report to the sick bay.

As all this was happening, 'Tibby' Brittain was pacing up and down the gym, watching the proceedings and keeping his eagle eye open for any infringement of the expected standard.

On one of his 'patrols', he reached the end of the gym and turned to start back. Suddenly, he let out one of his full-throated roars, guaranteed to shatter glass and terrify anyone below the rank of full Colonel. He swung round and raced for the door, waving his pace stick above his head.

The Trained Soldier under inspection froze in mid-cough and the MO dropped what he was doing and turned, as all did, to face whatever emergency the RSM had detected. All they saw was the rear of the British Army's senior warrant officer disappearing through the gym door.

Baffled, they swung their gaze around the gym. Then someone noticed the large round window, high in the wall above the entrance. Framed in this were three laughing NAAFI girls who had obviously been viewing the medical inspection from the building opposite, as a welcome and somewhat *risqué* break in their routine.

Pandemonium broke out. Under the MO's eye, trousers were hastily hauled up. Then the MO's team left hurriedly. The girls, sensing the end of the show and no doubt aware of the RSM's furious approach, disappeared quickly. Meantime, everyone else — Trained Soldiers, Sergeants-in-Waiting and orderlies — dissolved into roars of helpless laughter at the thought of Brittain's outraged attempt to confront the NAAFI girls.

A wag mimicked the RSM's standard cry when ordering an NCO to take some unfortunate's name — 'I never saw anything like it in all me life!!' And that caused another highly undisciplined uproar. Eventually, someone suggested it would be wise to return to barrack rooms before a furious RSM returned, and the parade broke up without being dismissed.

As they left, one of the laughing Sergeants told the Trained Soldiers: 'You think he's an unfeeling old bastard, yet here he is protecting your outraged modesty!'

———————oOo———————

Guards' Humour

'Don't get ailments above your station!'

Along with the Guardroom, the Caterham Depot Hospital was one of the places recruits most hoped to avoid. For one thing, something really had to be wrong with you before admission to its hygienic and antiseptic atmosphere was granted.

Chancers had no chance.

Despite, or perhaps because of its strict policy of nil-tolerance towards malingerers, the hospital was a rich source of Depot anecdotes. This one first appeared in the Household Division Magazine. It may contain some not-so-poetic licence!

On entering the hospital, one was immediately impressed by the orderliness and precision of the ward. The reflection of the 40-watt bulb glistened on the highly polished floor. Physiotherapy had not been introduced to the Guards Depot and the treatment for strained back muscles, arms or legs was ten minutes on the 'bumper', a heavy weight covered with bristles fixed to a six-foot pole which the patient undergoing treatment pushed and pulled up and down the floor to achieve a high degree of shine.

The blankets on the beds were folded like an envelope, all the beds were in line and a strong smell of carbolic soap and disinfectant filled the air.

The highlight of each day was Matron's rounds. Patients and orderlies experienced a weakening of the anal nervous system when anticipating her arrival. Instant paralysis occurred when she appeared in the ward doorway. One glance from her, it was said, could draw blood. She was not a vision of beauty, more like a granite statue, starched from head to foot. Her ample bosom, like the buffers of the Royal Scot, supported three rows of medal ribbons — enough to do justice to any Major-General. Even the dying lay to attention as she approached.

'What's the matter with you,' she asked Corporal Jan Hooper of the Coldstream.

'Haemorrhoids, Ma'am,' Jan replied, thumbs down the seams of his pyjama trousers.

'Haemorrhoids?' she screamed. 'Haemorrhoids? No, Corporal, you haven't got haemorrhoids — only officers have haemorrhoids, what you've got is piles!'

———————oOo———————

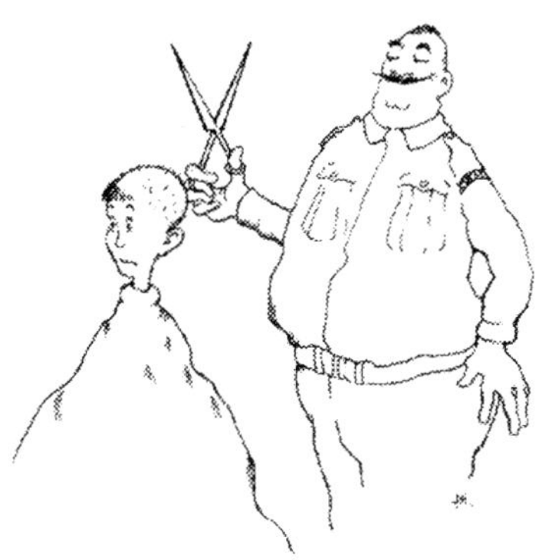

The Brigade Squads

The Brigade Squads

To be considered for a commission in the Guards, you have to prove two things above all: leadership potential, and the ability to match the physical endurance of men who have been tested in one of the military world's most unforgiving basic training regimes.

That's why the system of Brigade Squads was set up.

Many officers of the five regiments were put through their induction training at Caterham. Some potential Household Cavalry officers were among them. There were no privileges of potential rank. They went through the same gruelling programme of drill, weapon training, PT, and all the associated subjects which filled the days of the trainee recruit destined for the 'other ranks'. And they were put through this demanding course *by* 'other ranks' — squad instructors, drill sergeants, weapons experts, physical training and unarmed combat specialists — who represented the men they aspired to lead. Even the Trained Soldiers, the Guardsmen who supervised their barracks-bound evenings, commanded respect and instant obedience.

In these pages are recollections of some of those whose commissions were founded on a Caterham Depot base ...

The NCOs and WOs Who Shaped the Army

Lieutenant-General Sir David Scott-Barrett, KBE MC, was GOC Scotland, Governor of Edinburgh Castle, and Chairman of the Army Cadet Force Association in the closing years of an active and distinguished military career which began when he was commissioned in the Scots Guards in 1941. He saw active service in NW Europe in WW2, mainly with the Scots Guards' 3rd Armoured Battalion and was a Company Commander with the 2nd Battalion in Malaya. He held senior post-war appointments in Germany with BAOR, including that of General Officer Commanding, Berlin, 1973-75.

His memories of Caterham are from two spells at the Depot; first, in 1947, and again in 1959 when he was Second-in-Command and responsible for planning the move across Surrey to the new Depot at Pirbright, where he later became Commandant.

"The training system so superbly run behind Caterham's barracks walls produced dedicated, disciplined and motivated soldiers — the finest in the world. That, in itself, was a magnificent achievement. But it did more. It created a long line of senior Warrant Officers and Sergeants who influenced the whole British Army. That influence was exerted on officer recruits at Sandhurst, Mons Barracks, Aldershot, and many other establishments, and was responsible in no small measure for the standards of excellence and achievement which were the astonished envy of armies throughout the world.

This is no idle claim. The Guards' achievements and the reputation of Caterham was such that officers — selected to become future commanders of military forces of other countries — were placed in the charge of Guards' Warrant Officers and NCOs in order to absorb something of the Depot's training standards. I will never forget the stunned and amazed members of a Canadian officer cadre when they came under the ferocious control of RSM Freddie Archer!

The Depot system was unique. I was Staff Captain at the Depot in 1947, and saw at first hand the processes, refined over centuries to instil team work, the comradeship and pride Guardsmen have in their own regiments and the camaraderie which bonds them as members of the Guards' Division. Every young Squad Instructor and Trained Soldier knew that their reputations rested on the results achieved by the recruits in their charge and the system — skilled training, example, and hard discipline, leavened with morale-building humour — produced the results I have described. I never met a Guardsman who left the Depot with a chip on his shoulder.

*

Memories of Caterham Depot have stayed with me throughout my life and I have never swerved from the view that it was the finest military training school in the world."

A group of the Depot's Officers and NCOs in 1917. The military moustache was very popular.

Welsh Guards and Wasps

Lord Avebury celebrated his 21st birthday as a Welsh Guardsman at Caterham. As Eric Lubbock, he was commissioned a Lieutenant in 1949, but left the regiment in 1951 to enter politics. He famously fought and won Orpington for the Liberals in 1962, and was Liberal Whip in the Commons for seven years from 1963. His major interests include conservation, race relations, standards of conduct in public life, and human rights. He succeeded to the Peerage in 1971.

"I went to Caterham in September, 1949, after leaving Oxford with an engineering degree. The Brigade Squad was a bit of a contrast with the free and easy life of an undergraduate, but its novelty made it just palatable. There were a few others doing their military service after university, but most were straight out of school.

We were a motley lot and our ability to adapt to a military life was equally varied. Under the vigilant eyes of Trained Soldier Curley, we learned the arts of polishing boots and brass, and of arranging kit in a certain order for inspection — and survived the NAAFI tea.

On the parade ground, Sergeant Leitch was our master. He rapidly identified those who had problems distinguishing left from right, and those miserable creatures he labelled 'eggs'. When someone made a mistake, Sgt Leitch would come up to his victim and shout in his ear: 'You are an egg. What are you?' The miscreant had to reply, 'I'm an egg, Sergeant'. If the error was particularly grave, the Sergeant would vary the formula: 'You are Egg Number One. What are you?' and the recruit would have to repeat, 'I'm Egg Number One, Sergeant.'

Sometimes Sergeant Leitch would begin the morning inspection without having previously uttered a word to anybody. He would pass down the front rank, minutely examining every piece of kit, until he came to someone who was not quite up to standard. Then he would let loose with hyperbole: 'Your brasses are absolutely filthy,' he would declaim.

He had various ways of indicating his dissatisfaction with performance at drill. 'Next time you make a mistake like that, Laddie, you'll bounce' was one I remember, and the threat of the guardroom was frequently invoked though seldom used.

It was a hot September, and Caterham was suffering from a plague of wasps. Sergeant Leitch was as firm with them as he was with the squad. He would brandish his pace stick and shout, 'F... off, you bloody wasps' as if they were incompetent recruits. One day, we compiled a list of Sergeant Leitch's *bon mots*, and had a sweepstake which of them he would deploy first, next morning. As soon as he uttered the imprecation against the wasps, the whole squad burst into uncontrollable laughter. We all lost our names, but he never found out about the sweepstake.

My 21st birthday was a normal day at Caterham — I narrowly avoided losing my name. I am naturally too untidy and disorganised to have made a good soldier, but looking back on the experience, it may have done something to correct these innate faults.

*

The qualities of order and discipline which are exemplified by the Brigade of Guards are not widely valued to-day, at some cost to society. The experience of ten weeks at Caterham did nobody any harm, and most benefited from it."

In the 1800s

Seems wasps were taken seriously at Caterham. Someone recorded this 'daring deed of the Picquet Officer' in defeating a wasp invasion in September, 1887. Hopefully the unknown artist exaggerated just a bit . . .

Square-bashing & Fast Reactions

George Henry Hubert Lascelles, 7th Earl of Harewood and 26th in Succession to the Throne, served in WW2 as a Captain in the Grenadiers, being wounded and taken prisoner in Italy in 1944. He has a distinguished record in the world of opera and music and is a former director of the Royal Opera House, Covent Garden, and of the Edinburgh International Festival. He has been a BBC Governor, Chancellor of York University, Member of the Arts Council, and President of the English Football Association.

He remembers Caterham Depot ...

"Someone long ago told me that, as you go through life, each initiation to a new environment is a little less terrifying than the last. From the sheer purgatory of starting at your first school, you pass to the lesser horror of its successor and progress to a first job, or go to university, with hostility each time slightly less apparent.

Arrival at Caterham was certainly one of the most traumatic. This might be the real world.

Nobody was horrible to you in so many words. They just waited to pounce on mistakes. The yell which greeted the slightest infringement of military 'etiquette' suggested no possibility of apology or amendment. It demanded instant obedience and suggested further transgression would incur a fate not too far from death. As any repetition of the earlier mistake was unthinkable, death remained a reasonably distant prospect.

I suppose my recurring memory of Caterham is of incessant square-bashing — up and down the parade ground at what was often a slightly uncomfortable speed, interrupted by over-frequent about-turns which, if repeated by an impatient NCO, caused chaos and a bawling-out which made the reprimands of civilian life seem positively welcoming.

The consequence of incessant exercise on the barrack square was a ravenous appetite. One of the advantages of joining the Army in wartime was that you had excellent rations — not what you'd call imaginatively cooked but, in the hungry context, highly satisfying.

Life, or your enjoyment of it, was partly conditioned by an institution known as the Trained Soldier, a senior and possibly wily man who lived in your hut, was held in awe by all its inhabitants, and not only knew the rules himself but was there to ensure that you got to know them as rapidly as possible. He was the judge of whether your bedspace was tidy, your boots properly cleaned. He taught you how to lay out your kit for weekly inspection, and to yell out your regimental number when the officer came round.

I still put down 2622986 as part of my Lottery bid each week (it never wins anything, though).

Discipline, I afterwards came to see, was by far the most important thing we learned at Caterham, and for that I was unequivocally grateful. It was a product, probably, of these endless hours of responding to barked orders on the parade ground. You acquired the habit of immediate and, just as important, unquestioning obedience. If you were marching along faster than you would choose (but choice wasn't on the agenda) and someone yelled 'ABOUT TURN!', you didn't ask 'Why?' In the context of battle, the same immediate and instinctive reaction was to be pretty useful.

I suppose the two winter months spent at Caterham discovering the Army were, on the whole, not unenjoyable. I doubt, though, if that would have been my summary at the time ... !"

Invasion Threats and Justice

Christopher Thursby-Pelham was 18 when he joined a Brigade Squad at Caterham in 1940. Six months later, he left the Royal Military College, Sandhurst, as Second Lieutenant Thursby-Pelham, Welsh Guards. He saw active service with the 3rd Battalion Welsh Guards in North Africa and Italy. He commanded No 8 Company in 1954-56, and returned to the Depot as Commandant 1960-63, by which time the Depot had moved to Pirbright. He commanded the Welsh Guards 1964-67 and left the Army in 1976 with the rank of Brigadier, having been Chief-of-Staff, London District 1969-72.

"It was late one afternoon in September, 1940, when I reached Caterham. I was squadded with other Scots, Irish and Welsh Guards recruits in a hut in the Irish Guards lines, near Green Lane. I was glad to discover I knew one or two of the others. During our stay at the Depot, we were to be given over to Sergeant Tom McCart and Trained Soldier Stoneham to be converted into something resembling Guardsmen.

The Battle of Britain was at its peak. We could have had a grandstand view if we hadn't been fully occupied with the demands of training, set for us relentlessly by instructors whose examples we had to aspire to match. However, we could see the vapour and exhaust trails and hear the sound of machine gun and cannon fire as Spitfires and Hurricanes engaged the German raiders, 20,000 feet above our heads.

On about the third Sunday, the Commandant addressed all newly-joined recruits in the Depot Chapel. I remember that he made two points most forcibly. Presumably addressed to those of us tempted to test the effect of uniform on the fair sex, the first was an absolute ban on taking ground-sheets with us when we were 'walking out'. The second point was a reminder that we were now members of the finest military brotherhood the world had ever known and that we had a great tradition to live up to. We were proud, of course, although there were times - when we were being chased round the square, for example — that something rather more easy-going seemed preferable.

I don't think many in our squad found Depot life too testing. Since the age of eight or nine, we had all spent nine months of every year at school away from home, so homesickness wasn't really a problem. Most of us, too, were pretty fit. Knowing I was bound for the Army, I had taken the precaution of working in a lumber camp in the Forest of Dean that summer, and that gave me a running start on the physical side of the training.

The sounds of war and its impact were everywhere. When we marched to the gymnasium, we passed the shattered remains of the Welsh Guards barrack block which had received a direct hit by a bomb a few weeks earlier, killing eighteen recruits. Such calamities were not allowed to interfere with the Depot routine and we got a first sight of one of the strengths Guardsmen acquire from their training — the ability to continue to operate through fire and wreck.

Looking back, I suppose the principal lessons we learned were those of fortitude and justice — fortitude in the sense that nothing was allowed to deflect us from our immediate objective. For instance, I cannot recall a drill parade which wasn't interrupted by the air raid siren. With airborne invasion a constant threat, we had to fix bayonets and man the slit trenches round Fox Square. When the 'all clear' sounded, we returned to the square, unfixed bayonets, and carried on with our drill.

We learned about justice, in the sense that actions had their appropriate rewards. Credit was given where it was due, and punishment was awarded when deserved. Justice, as we experienced it, was firm but fair. Also, the virtues of temperance and prudence were sown in our minds and we cultivated them — with varying degrees of success — as time went on.

In our mixed squad, we learned to appreciate the special brand of humour which is the distinguishing mark of the Micks.

'Keep your caps on! Don't you know there's woodpeckers about?' when a gust of wind blew off most of the squad's headgear on a drill parade. And to a recruit whose morning shave was just short of winning him a punishment drill: 'Next time, stand closer to the razor and take the safety catch off!'

Often, we disobeyed the rule against talking after lights-out. Trained Soldier Stoneham admonished us, but it still happened. Some of our stories must have shocked him. One night, he got out of bed and knelt beside it. He told us, 'I was praying for you!'

By the time we left Caterham for the OCTU at Sandhurst, we were already proud to be Guardsmen and, with the assistance of Grenadier Company Sergeant Major John Lord, even prouder when we were commissioned as Ensigns in our respective regiments four months later.

We met our much respected Trained Soldier Stoneham just once more. Four of us were driving up to London from the Training Battalion at Sandown Park. Coming down Putney Hill in my open Vauxhall, we saw him standing outside the pub at the cross-roads. I pulled up with a squeal of brakes and we leaped out and rushed over to him, yelling: 'Trained Soldier! How are you?' 'Glory be!' he gasped. 'I thought you were from the War Office!'

Sadly, I am the only surviving member of that happy reunion."

The Hut Parade . . . and a Discovery

Richard Henry Piers Butler, 7th Viscount Mountgarret, went through Caterham Depot in its final Brigade Squad. He was commissioned in the Irish Guards in 1957, and retired with the rank of Captain in 1964. A keen sportsman, he was President of Yorkshire Cricket Club from 1984 to 1990. He contributes this memory of Brigade Squad life under a hard, but human Sergeant instructor.

"Our Squad Instructor was Sergeant McGowan. He was Scots Guards. He was as hard as they come. He was terrifying. And as recruits, we hated him from the first.

To us, he was a demon who chased us mercilessly when he judged our progress as not up to the standard he was expecting. He introduced us to marching in treble-quick time as a means of making his point, and had us on charges — particularly in relation to lapses in cleanliness of our kit. We thought him thoroughly bloody-minded, and we loathed him.

One day, about half-way through the course, it came on to snow at night, and snow indeed it did. In the morning, there must have been upwards of nine inches on the ground. Our first parade that day was to have been drill on the square, and we were laughing among ourselves, knowing how impossible it would be to have any drill in such circumstances.

I should say at this point that we lived in a Nissen hut with a wooden floor and windows that hardly worked. For heating, there were two stoves in the centre of the room, about a third of the way down from either end. As we were the last Brigade Squad, the other four huts usually occupied by officer recruits were empty.

We were enjoying our thoughts of a snowy reprieve from drill when the door swung open. The snow blew in, and Sergeant McGowan strode in, straight as a ramrod, impeccably turned out as always, and with evil in his eyes.

'Now,' he barked, 'you lot think that just because of a little snow, there will be no drill parade. Well, I have news for you. In one minute you will all be in the next-door hut, properly turned out, with not a suggestion of snow on your boots, and we will have our drill in there. Last man in will be confined to barracks. MOVE!' We grabbed our rifles and fled.

The drill began in double-quick time, up and down the hut which could not have been much more than twenty yards long. 'About turn, left turn, right turn, about turn, right turn, left turn,' bellowed Sergeant McGowan. Round and round the stoves we marched, trying to avoid them, but the speed at which we were spun around in all directions proved too much. Finally and inevitably, some of us collided with the stoves, bringing them and their chimney pipes crashing to the floor.

We collapsed into laughter, but on recovering a degree of composure we noticed the fierce features of Sergeant McGowan melting before our eyes. A smile crossed his face and, not being able to contain himself with the absurdity of the situation, burst into laughter himself.

This blink in the hard and constant discipline was a watershed in that it gave us a new perspective on our training. Sergeant McGowan, we not only saw as a firm, skilled and demanding instructor, but combined with these qualities was the essential degree of humanity which, in itself, is a basis of leadership.

The atmosphere changed and we actually looked forward to drill parades and we did our best. We were marched at the approved Brigade pace and could feel ourselves improving steadily, as individuals and as a squad. He was as correct as ever, but our initial opinions of him had been shattered.

By the end of our time at Caterham, no man could have commanded more respect, and we parted firm friends.

*

It was there I came to recognise the unique qualities of Non-commissioned Officers in the British Army, and in the Brigade of Guards in particular."

All Ranks — We Were All Guardsmen

Captain John AJ Powell served in the Welsh Guards from 1944 till 1947. He now lives in Plympton, South Australia, but his memories of Caterham are still vivid. He contributed this thought:

"Napoleon once said that an Army never has bad soldiers, only bad officers! The converse could also apply; first-class soldiers have first-class officers.

I always used to think that the backbone of the Brigade of Guards was formed by its NCOs. Only later did I realise that the perfection of the Guards is due to a perfect combination of all ranks, Guardsmen, NCOs and Officers. What a damned good combination! I am proud to have been part of it.

*

And the foundation on which it was built, the mould which cast the efficiency and reputation of the elite regiments of the Brigade for a hundred years was **The Guards' Depot, Caterham.**"

In the 1800s

One of the awkward squad gets the Sergeant's view of his "military bearing" — 'You are standing just like an old cab 'orse.'

The Padre's Story

The natural optimism of the human spirit led most servicemen and women into a hopeful expectation that they would survive WW2. Among many 18 year-old conscripts, however, joining a Guards regiment at Caterham meant putting firm ideas of a future civilian career on hold till 'this lot' was sorted out. So it was for 2623839 Recruit Huggins J. when he arrived at the Depot in January, 1943. The war in Europe was about to enter its most hazardous phase and life's perceptions had shrunk to the immediate pressures of having to become a Grenadier Guardsman and join an active service battalion bound for the invasion of Europe.

Five decades later, he has retired as The Rev. JWH Huggins and lives with his wife in Frome, Somerset. He holds the honorary position of padre to the Bath branch of his regimental association.

"It's funny how little and inconsequential memories persist over the years. I remember CSM 'Wacky' Jones of the Welsh Guards forcefully instructing our Brigade Squad on the parade ground that our cap badges must LOOK LIKE SILVER! He also told us to use something called 'Heelball' on our rifle butts, despite being told repeatedly by Sergeant Llewellyn that there was none in the stores. My first Caterham haircut stands out, of course. I have always been keen on drawing and did a caricature of the barber while waiting my turn. Someone must have shown it to him. When I sat in the chair, he almost shaved my head, leaving the merest tuft in front!

I tried on my issue greatcoat. The material was very stiff and stuck out on either side like wings. 'Look out! You're going to take off', was among the ribald comments I received - and one of the milder things said to me in my early days as a soldier.

A crisis situation: the Adjutant was due in minutes to make an inspection and my kit was nowhere near ready. My friend, Gilbert Lamb (later Major GW Lamb) who had the next bed, told me to tidy myself up and said he would see to my kit. Unlike me, he was immensely efficient and in no time he had everything in perfect order, right down to the last detail. A minute or two later, we were standing rigidly by our beds for inspection - and I was given credit for the high standard I had achieved! And, of course, Gilbert Lamb and his kit were entirely overlooked.

My Brigade Squad was instructed by the Welsh Guards, but our Weapon Training instructor was Sergeant Golightly of the Scots Guards. One day, he took a small group of young potential officers (in other words, raw recruits) to the local pub and showed us how to drink beer from a flower vase. I had six pints and only just managed to get back to barracks. At the weapon training session next morning, Sergeant Golightly appeared in his best battledress uniform. He explained: 'I couldn't have been myself last night. I put my underpants down to crease and threw my trousers on the floor!' But that isn't *exactly* how he put it."

Lieutenant John Huggins joined the Grenadiers' 1st Battalion in Yorkshire in 1943 and took part in the invasion of Europe in June, 1944.

He was wounded a month after landing in France, was operated on in England and rejoined his battalion that September. He saw active service in France, Holland and Germany as a platoon commander.

A night reconnaissance just before the Rhine crossing in November, 1944, turned out to be the most terrifying experience of Huggins' life. Sent out with a Guardsman to discover if the village of Schummenquartier was still occupied by the enemy, the pair crept close in to watch and listen. There were no lights or sounds. Then Huggins found that his sweater had caught on the prongs of an 'S' mine - they were on the edge of a minefield.

Very slowly and with extreme care, he detached his sweater, then located the direction of the mine's trip wire by running his fingers gingerly along it. They managed to step back over the wire and drew back from the minefield. Then they fired a Bren and two-inch mortar into the village. There was no reaction.

Next night, the Germans sent out a patrol to find out what had been going on. It was ambushed by Huggins' unit and one of the prisoners they took confirmed that the village was strongly held.

Returning from another night recce, Huggins took a wrong compass bearing and lost the way.

'Eventually, we came to the outer perimeter of one of our companies and were challenged. At a distance, we proved who we were and survived only by naming a few characters, such as the company cook — and using some strong language! We were lucky. That company had machine guns trained on fixed lines at the place where we were standing.'

When the war in Europe was won, he was posted to the 3rd Battalion and went to Palestine where he remained till his demob with the rank of Captain in May, 1947.

He was ordained in 1951 and began his ministry in Northumberland.

Today, he comments: 'The Guards' training is valuable. With its discipline and severe training, it strengthens character. For that, and for the friendships I made, I look back with gratitude.'

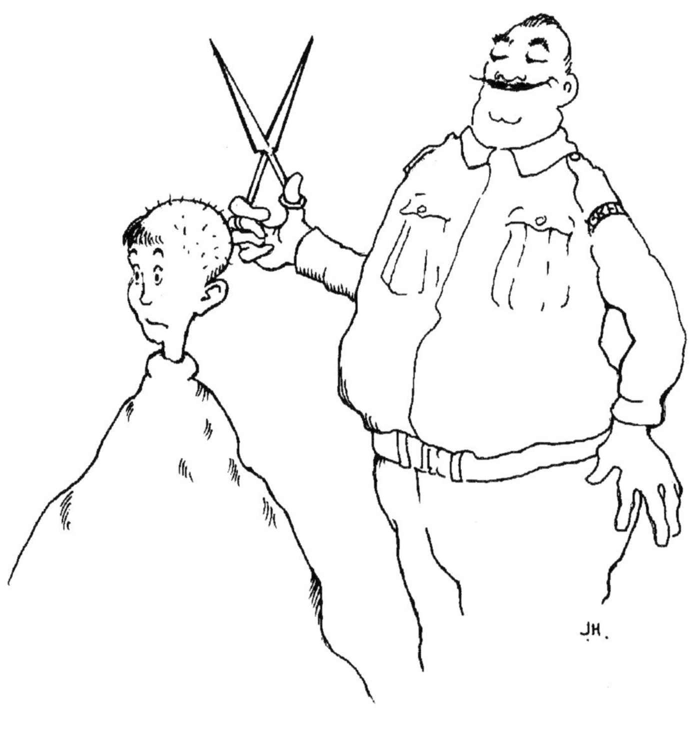

The Huggins Haircut — as he remembers it!

Mafeking - And Up, Up And Away!

The Relief of Mafeking in May 1900 produced an explosion of national rejoicing when news came through that troops had broken the seven-month Boer siege of the city. But while the whole country celebrated, a bizarre coincidence gave the victory special significance at Caterham. With the relieving force was a Scots Guards officer who had been Adjutant at the Depot nine years earlier. He was Major Fletcher Baden-Powell, and the embattled garrison his force relieved was commanded by the 'Hero of Mafeking' his older brother, Sir Robert Baden-Powell, who later founded the Boy Scout and Girl Guide movements.

Fletcher Baden-Powell was 22 when he joined the Scots Guards in 1882. He served with the Guards' Camel Corps on the Nile in 1884-85, then was ADC to the Governor of Queensland, and a member of a punitive expedition in New Guinea before taking up the post of Adjutant at Caterham in 1891. That he was an extremely brave and adventurous young man cannot be doubted. He served with his regiment throughout the Boer War, taking part in seven major actions, and returned from the reserve to join the 2nd Battalion in France and Belgium in WW1. With the 2nd, he took part in the key attack on the Hohenzollern Redoubt. Baden-Powell was appointed Divisional Bombing Officer to the Guards Division and was also recognised widely as an expert in the art of camouflage. But his daring and adventurous spirit is best illustrated by his earlier career. In 1884, he invented man-lifting kites as an early form of aerial observation and reconnaissance. He himself made frequent ascents by kite during the Boer War. His fascination with flight led him to revive the Aeronautical Society in 1897 and he later became its president. He flew with Wilbur Wright not long after the Kitty Hawk break-through, thus becoming one of the first men to experience powered flight as a passenger. An enthusiastic balloonist and glider, he took to the sky in another form of flight in 1913 - by Zeppelin. He also found time to write books on warfare, ballooning, and practical aeronautics. These are the formally recorded facts of Baden-Powell's career. They reveal little, however, of the personal man. For a glimpse of him 'off duty' we are indebted to the Croydon Chronicle of May 1st, 1891 . . .

MUSICAL AND DRAMATIC ENTERTAINMENT AT THE GUARDS' DEPOT

On Thursday evening the Theatre at the Guards' Depot was literally crammed on the occasion of a musical and dramatic entertainment promoted by Mrs. Colville, Colonel Colville (the Commandant), and officers of' The Depot. That these entertainments have lost none of their popularity was demonstrated by the large audience which included all the officers and a considerable number of the non-commissioned officers and men besides a large sprinkling of civilians among whom were many leading residents of the neighbourhood. The arrangements for the evening were carried under the direction of Captain Gunton, by Staff Sergeants Darkin and Mackenzie, Corporal Johnson, Private Taylor and others; and from first to last the performance went without a hitch.

The first part of the entertainment was commenced by Corporal Johnson, Grenadier Guards, singing a new song entitled 'Comrades' which proved most attractive to

Tommy Atkins by whom it was loudly re-demanded. Staff Sergeant Mackenzie followed with the grand song 'True till death' which he rendered with telling effect and fully deserved the loud applause which was awarded to him . .

. . . the second part of the performance consisted of the old, but ever popular farce, *Turn Him Out !* which was most admirably mounted. and played with the following cast- *Nicodemus Nobbs:* Lt. Macdonald, *Mr Mackintosh Moke:* Lt. Baden-Powell; *Mr Eglantine Roseleaf:* Sgt. Ponsonby; *Julia:* Miss Bessie Crossley; *Susan:* Mrs. Colville.

To say that the farce was well played would he poor praise, for it was rattled through in a style worthy of an Adelphi company in the old days. Lt. Baden-Powell, the popular Adjutant, fairly astonished his most immediate friends by the lively manner in which he sustained the part of *Mr Mackintosh Moke*, as in everyday life he is a gentleman of such quiet manners. He fully rose to the occasion, however, and as the indignant husband, ruthlessly turned out of his own house through a mistake. He showed dramatic abilities of a very high order, and which elicited for him warm applause.

Scots Guards

Fantasy from the Trenches

The letters a young Guards officer wrote to his children from the trenches of WW1 have charmed generations of youngsters who have read the books and seen the film the letters inspired.

To-day, the Dr Dolittle stories are still recognised as children's classics.

Their author, Hugh Lofting, was born in Berkshire and studied civil engineering at the Massachusetts Institute of Technology and the London Polytechnic. His career took him prospecting and surveying in Canada, followed by years of civil engineering on railway projects in Africa and Cuba. By 1912 he had decided on a career as an author and settled in New York.

He was 28 when war came. He returned to Britain, went through Caterham and was commissioned in the Irish Guards.

As a mental escape from the appalling conditions of the trenches in France and Belgium, Lofting began a series of story-letters to his children, constructed around Dr Dolittle, a gentle eccentric who has learned to talk the animals' languages in order to treat them more successfully.

Lieutenant Lofting sustained severe wounds and was invalided out of the Guards.

Back in civilian life, he worked as writer and artist for the British Ministry of Information in New York. In his spare time, he developed the stories he had written for his children and the first book, The Story of Dr Dolittle, was published in 1920. It was a runaway success. The public response was so positive that he published a Dr Dolittle book annually for seven years, one winning the award for the best children's book of the year. Many of the illustrations were drawings he had done himself.

The film, Dr Dolittle, was produced in 1967 and starred Rex Harrison as the animals' physician in Puddleby-on-the-Marsh. The screen version renewed widespread interest in the Dolittle series and the books are still in demand in the public libraries.

*

Although he lived in America after the war, Hugh Lofting's stories were set in England and are recognised internationally as children's classics. He died in 1947 in Santa Monica, California.

———————oOo———————

Guards' Humour

The late King Hussein distinguished himself as an officer cadet at Sandhurst, where he came under the care of senior Guards instructors, neither seeking nor receiving any privileges to reflect his royal status. The disciplinary code at the college was leavened, however, by the humour which is characteristic of Guards' training. On parade, 6ft. 4ins. Grenadier Company Sergeant Major Victor King paused before the 5ft. 4ins. King Hussein and growled:

'King Hussein, Sir, there are two kings on parade - you, Sir, and me, Sir - and we both know who's the more important King . . . Don't we, Sir!'

And everyone on parade that day, not least CSM King, knew the answer.

———————oOo———————

The High Fliers who Limped

'Amazing!' was how Depot Commandant, Lieut-Colonel VF Erskine-Crum, CIE. MC. described the Brigade Squad of potential officers which went through its Depot training early in 1959. The description was well-earned.

It was also prophetic.

Not only did the squad excel under Superintendent Sergeant Peter Horsfall, in later years their members all became high achievers, including a Deputy Prime Minister, two Peers of the Realm, a Brigadier who commanded the 1st Battalion, Welsh Guards in the Falklands, a Colonel in the Scots Guards now serving as a Gentleman-at-Arms, and a Director of a large Gloucestershire estate who is also an officer in the Queen's Bodyguard of the Yeomen of the Guard.

Astonishingly, in the light of the Depot performance it was to achieve, the squad got off to a very bad start, recalls their chief instructor who retired as Major Peter Horsfall MBE, Coldstream Guards — and then served as Staff Superintendent in the House of Lords for fifteen years . . .

"The physical training instructors had established a practice of taking recruits over the obstacle course in their first week-end at the Depot 'to improve their fitness'. The result was the opposite. At the time, the ground was frozen hard and a third of the recruits in the two squads involved suffered ankle, leg and foot injuries. As a result of this and other casualties, the squads were formed into one and, unhappily, Sergeant Rudd, the other instructor, was returned to normal duties with his Scots Guards company. The PT instructors were highly unpopular with the Squad Officer, Captain Trevor Dawson, Scots Guards, and Administrative Sergeant George Nash of the Grenadiers.

Assisting with the training were two outstanding weapon training instructors, Sergeants Tait and Mason; a first-class PT instructor, Sergeant Stokes; and two terrific Trained Soldiers, J. Fisher and B. Thorpe. Backed up by 'Perry' Mason and 'Spud' Tait, who finished as Sergeants, and Derek Stokes, who ended up a CSM, we made up a first-class training team.

The recruits we were left with were J. Ashton, R. Bruce, C. Constable Maxwell, T. Cotterall, M. Heseltine, D. Jarratt, P. Johnson, C. Lawrence, C. Madden, H. Myddleton, A. Pearson, J. Rees, J. Rickett, M. Rush, R. Saunders, The Hon S. Scott, The Hon J. Skeffington, The Hon. T. Tollemache, M. Wait, J. Willis and T. Wills.

There can be no greater honour for a non-commissioned officer than to carry out the initial training for those destined to be officers in his own, and other regiments. He must put out of his mind that in a few months the young men he is training will be commanding him in certain situations. At times in the Depot, the instructor was tougher on these than he would have been on other recruits, but this was because Brigade Squads had only eight weeks to achieve the same standard as other squads reached in twelve. No easy task.

The squad was divided between the two Trained Soldiers in two very old-fashioned huts with primitive ablutions. The day consisted of nine periods of 45 minutes, with 15 minutes between periods. 'Horsfall's Flyers', as the squad was known throughout the Depot, got the nickname because of the speed they were marched around barracks and the fast clothing changes they had to make between periods.

At the two-hour shining parades each evening, each recruit had to give five-minute lecturettes to his mates on different subjects, in addition to cleaning his kit. Recruit Heseltine excelled at these. His talks were always interesting and I sometimes allowed him, and some others, to go on longer than the allotted five minutes. This was to prepare the recruits for the ordeal of the Regular Commissions Board in Westbury.

Recruits were not allowed to walk out of barracks until after passing off the square at the fourth-week inspection by the Adjutant. This consisted of foot drill, picquet drill and saluting collectively and individually. The squad did well — a minor miracle because several recruits had missed numerous drill periods due to the injuries suffered on the obstacle course in that first week-end. A factor was the immaculate state of their kit, thanks to the hard work of Trained Soldiers Thorpe and Fisher. Both led by example and their own kit was of the highest order. Years later, I learned that it was common practice for squad members to be still cleaning kit well after the 2215 hours Lights Out — in the ablutions or toilets.

The fourth week inspection over, four Squad Leaders were to be chosen from among the squad members. There were six really good contenders but in the event, alphabetically, the choice was Bruce, Heseltine, Madden and Wills. If Heseltine had not missed so much drill, due to his leg injury, I believe he would have been the clear front runner in these appointments.

When the squad deserved it, I chased them as a punishment. I did not believe in taking recruits' names and marching them in front of the Squad Officer. Not only did I think this wasted valuable training time, I didn't believe it helped a potential officer to have black marks on his record so early in his service. The nearest I came to this was when one Trained Soldier found a brass knuckle-duster in a recruit's kit. I was furious and found it hard to believe a future officer could behave so foolishly. If Captain Dawson had found out, I'm sure the recruit would have been kicked off the course — despite being a good natural leader.

If the squad were idle, I marched them down to the old deserted NAAFI in the Fox Lines. There, I would drill them in very quick time until satisfied that they were all working for me. As the chasing went on, I dismissed them one at a time when I thought they were up to standard. I was not proud of this tactic but knew that it was necessary if they were going to pass out with flying colours. I have to report that almost always the last recruit to be dismissed on these occasions was a certain Coldstream. He became a good officer!

One night towards the end of the course, my wife and I were in our married quarters in C Block and heard a fracas outside. It was late and those responsible had obviously been out on the town and had sunk a few. I told Mary I would sort them out. To my surprise and embarrassment, I found they were members of my Brigade Squad! Instead of 'sorting them out', I took them in and Mary made them tea and sandwiches.

To this day, those involved recall their astonishment when they realised that Sergeant Horsfall was human. As to having a wife and children — NEVER!!

The squad's passing-out parade in the eighth week was a joy. Thorpe and Fisher had done their stuff and the Commandant, Colonel Erskine-Crum (who later became a General) told the squad theirs was the best turn-out he had ever seen. I didn't think the drill was all that good, but the Commandant and RSM 'Jolly' Roger, Scots Guards, obviously allowed for the shocking weather from January to March, the injuries on the obstacle course and the wonderful results achieved at the Regular Commissions Board. But who was I to grumble. It made my name.

Following the careers of the squad has been interesting to say the least. Michael Heseltine became Deputy Prime Minister, Brigadier Johnny Rickett commanded the 1st Battalion, Welsh Guards in the Falklands and has been an outstanding soldier, Timothy Tollemache is now a Lord, John Skeffington is now the Viscount Massereene and Ferrard, Peter Johnson was a Colonel in the Scots Guards and is serving as a Gentleman-at-Arms; Tom Wills runs a large estate in Gloucestershire and is an officer in the Queen's Bodyguard of the Yeomen of the Guard. I believe every member of that squad has been successful in his chosen career.

Sadly, Trevor Dawson, Ashton, Jarratt and Rush are dead. The survivors still attend reunions, the latest on the 40th anniversary of the squad's formation, in January 1999.

It was a privilege to be part of an outstanding team which contributed to the training of so many fine young men. Although some of those in the squad served in their regiments for a relatively short time, they are all emphatic that it helped to shape their lives in large measure. It must be a tribute to the old Guards' Depot at Caterham."

A Senior officer's over-the-top fears about Bohemian influences on young officer intakes.

The Depot in WW2

An Irish Example

The reputation of Captain Desmond Lynch, DCM, was earned in the ranks and as an RSM who got the best out of his men by setting a startling example in all aspects of soldiering, particularly courage.

An officer who served with him in the Irish Guards during WW2 recalled: 'He had a reputation for ferocity which was unequalled in the Micks. All I can say is, thank God Lynch was on our side!'

He won the Distinguished Conduct Medal in April, 1943, in North Africa. He was platoon sergeant in the leading company during an attack which sustained considerable casualties. His company commander and two platoon commanders were killed or wounded, and the company sergeant major killed. After laying his wounded platoon officer in a slit trench, Lynch saluted and requested: 'leave to carry on, Sir', then took charge of what was left of the company. Under his command, its remnants attacked and gained its objective. His small force was continually attacked for three days and dwindled further. But Lynch, who was half blinded by blast, moved a Bren gun to an advanced position to anticipate and beat back the main German infantry attack which followed. The son of a soldier, Desmond Lynch was born in Dublin, joined the Irish Guards in 1937 at Caterham and saw service in Egypt and Palestine in the following year. After a spell as instructor at the Royal Military College of Science, he returned to the Guards and rapidly gained promotion. At 6ft 3ins and heavily built, Lynch was always immaculate, with a strong Irish brogue, even in his words of command. Before the engagement which led to his DCM, he led his platoon in an attack on a German position, taking prisoner an artillery officer and two sergeant majors, and leaving most of the others dead. It was reported that the German officer wept uncontrollably and when asked why, replied that he was mourning his dead companions. Lynch urged him to be a man and accept things philosophically, advice which gave the German little consolation as Lynch had killed most of them himself.

Wounded twice in North Africa, Lynch nevertheless took part in the Anzio landings and the Italian campaign which followed.

Wounded again, he was taken prisoner, nursed by nuns in Rome, then moved to Stalag 7A in Germany. It was said that he escaped by strangling his guard.

After the war, he had postings in Europe and West Africa, then became the first drill sergeant at the All Arms Wing, Caterham. A spell at Pirbright followed, then three years as RSM by invitation at the Indian Military Academy, Dehra Dun.

Lynch proved himself tactful as well as resourceful when he did a tour of duty at Eaton Hall and Mons Officer Cadet School in the '50s. The adjutant had to drive in each day, over roads subject to frequent traffic delays, but RSM Lynch ensured that he was never late for parades by having the clock strike the hour when the adjutant had arrived. Appointed MBE in 1960, he was commissioned the following year and joined the Commonwealth Brigade at Taiping and Malacca in Malaya for three years. Later, he was adjutant of the Singapore Guard Regiment. Final posting before he retired in 1973 was to the Ministry of Defence as a staff captain.

Even in retirement, he was active. He was assistant adjutant and security officer at the Royal Military Academy, Sandhurst. He enforced regulations impartially. Breaking the 25-miles-per-hour speed limit meant the offender's car being banned for two weeks.

The Commandant's wife once incurred the full penalty herself. Despite his reputation, Lynch had a great sense of humour as well as a strong humanitarian streak. In retirement, he did much valuable work for charities and as a prison visitor.

When he died in 1997, aged 75, he was widely mourned in the Guards Division, and among many who had the privilege of knowing him in many other spheres of his activities.

—————oOo—————

Guards' Humour

The ramshackle fleet of little ships was bringing British and allied troops back from the chaos of Dunkirk, then under attack by German aircraft. A reporter, desperate for a first-hand view of conditions on the beaches, dashed up to a young Guards lieutenant as he stepped ashore from a crowded barge.

'Tell me what it's really like over there,' he urged.

The young subaltern smothered the first response which leaped to mind. 'My dear chap,' he replied languidly, 'You can't imagine - the noise, and the PEOPLE!'

—————oOo—————

In the 1800s . . .

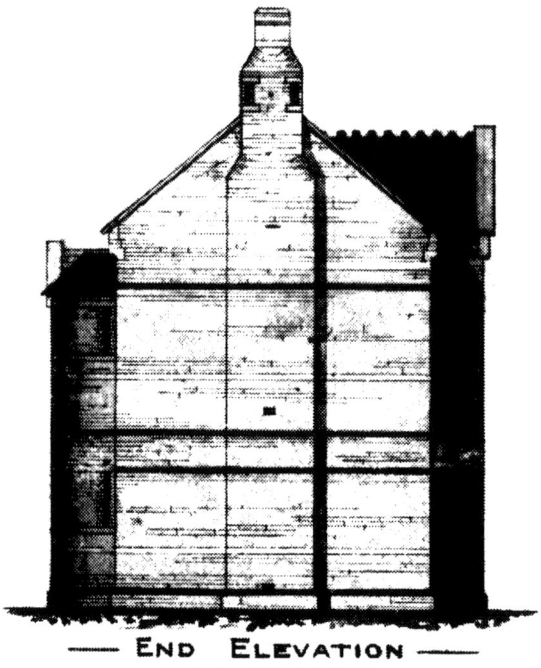

From the architect's (Major Sneddon) drawings of 1875..

Blitz Alarms

Caterham Guards' Depot under aerial attack are memories of WW2 by former Grenadier (2621112) Kenneth Timmis who did his recruit training in 1940. Lists of the dead and injured were lengthening and he came within an ace of becoming a casualty himself. He wrote from his home in Gerona, Spain.

"It was August, 1940, and I was in Corporal Kirby's squad. The war effort was building up and they were pushed for room. So many new recruits had to be accommodated that our squad were put in what had been the Depot museum. The floor, I remember, was like polished mahogany.

I was walking past the barrack blocks with another recruit one day when we heard an aircraft at low level. I had studied aircraft identification before I enlisted and recognised the 'plane at once as a twin-engined Junkers 88. I shouted a warning, but my mate laughed and claimed it was a Blenheim. Then bullets began to ricochet off the road in front and we dived into a doorway. As a bugler sounded the alarm, we ran to the shelters. Apparently, the only damage done in that hit-and-run raid was to the pay bloke's bike! We heard later that the Junkers had been shot down at Redhill.

During another daylight raid that August, I stopped at the entrance to one of the shelters and watched a mass dogfight in the sky. Strange, at that height they looked just like model aircraft. I could hear spent cartridge cases falling nearby. Then I heard the banshee shriek of dive bombers and took cover. The shelters were in a rectangle enclosing the cricket pitch and had been grassed over.

The noise of exploding bombs was deafening, one close to our shelter. We emerged when the all-clear sounded to find that one of the Coldstream shelters had received a direct hit, and a number of recruits killed. The bodies were wrapped in blankets and taken into the Depot chapel.

Almost every night, we heard the drone of German bombers on their way to attack London. Once, when we were doing our shining parade by candlelight and listening to the squadrons passing overhead, we heard the whistle of a bomb coming down and were quickly on the floor. We waited for the explosion, but heard nothing. An unexploded bomb, we thought, and got on with our tasks.

Daylight, however, exposed the full horror of the situation. On our way to the messroom for breakfast, armed with our mess tins and cutlery, we saw that Victoria Block had been practically destroyed. Bodies of dead Guardsmen had been laid out on the grass nearby. I remember the body of one poor chap was still hanging in the ruined building.

There were other alarms. One pitch-black and freezing night in September, we were about to retire when an officer and platoon sergeant came into the room. We were told in sombre terms that invasion was believed to be imminent. We must sleep in our battledress uniforms, and each man was issued with two hundred rounds of ammunition.

We were told that coaches were being lined up on the barrack square, ready to take us to the coast, if necessary. We spent a wakeful night, lying or sitting on our beds. Daylight came without incident, however, and there was a rush for the chap who brought the morning papers into the Depot, to see if the censors had allowed any reports about the night's alarms.

There was nothing, and we heard no more about it.

Much later, I returned to the Guards' Depot as a squad instructor. I trained a number of squads, the last in July, 1943, by which time I was a three-striper. One of my squads was considered the best in the Depot and was chosen to give morale-boosting demonstrations of foot and arms drill in various parts of London.

At that time, my CO was Captain H.S. Goodheart-Rendell MC, and the CSM was W. 'Scaly' Betts. The captain had retired, but rejoined on the outbreak of war. He was a great chap. He took the Depot staff to his home, Hatchlands, in Guildford — a beautiful place where he laid on beer and food. He sent me a Christmas card and short letter each year until he died. Hatchlands is now in the care of the National Trust.

On a lighter note: I was Sergeant of the Guard on the Depot main gate one day when an officer came round to check that everyone knew what their duties were in the event of emergency. He asked one of the Irish recruits, 'What would you do if the NAAFI caught fire?' The recruit knew. Without hesitation, he replied: 'Use the Church Army, Sir!'

These were difficult and dangerous days, but it wasn't all doom and gloom. Strangely, however, there was one other very unfunny incident in my Depot experience which made me roar with laughter years later.

I was a Corporal at the time, and I'd been out on a 2am pass. In fact, I got back to the guardroom at 12.10am, then found I had lost my pass. Things then went rapidly downhill because the Sergeant-in-Waiting had me down on the tattoo sheet as due back at 23.59pm. I explained the situation to an MP Sergeant we shall call Green. Without making any inquiries, he put me in a cell. When I was marched in before the Company Commander in the morning, the truth was established and the outcome was 'case explained'. It was a lousy experience.

Years after I left the Guards, and was a sergeant in the Worcestershire Police — now the West Mercia Force — a chief inspector friend invited me to a dinner at the Bramshill Police College, and laid on transport so that I didn't have to drive. I had had a few pints when I sat down to a lavish dinner. We were in dinner jackets and had been asked to wear miniature medal ribbons. Opposite me was a chief inspector from a force in the south.

He asked: 'What mob were you in?' When I replied, 'the Grenadiers', he asked if, by any chance, I had known his father who had been a police sergeant at the Depot, adding that his name was Green.

The memory of my night in the cells came flooding back and I saw red. I told the chief inspector: 'Your father was a bastard!'

Immediately, I realised that I could be in trouble. Instead of the feared reaction, however, the chief inspector just grinned.

'He still is!' he said.

Bombs and Bed Blunders

The Caterham area was among the districts most seriously affected by enemy action during the Battle of Britain.
Aerial dogfights daily pitted Spitfires and Hurricanes against the Messerschmitt escorts of the German bomber squadrons heading for London. High altitude vapour and exhaust trails of the opposing fighters wove deadly patterns across the blue skies of the Summer of 1940.

Many bombs fell in and around Caterham. Crashed aircraft, British and German became regular if temporary features of the countryside. There were two significant reasons for the close attention the German Luftwaffe paid to the town and its immediate area.

The first was Kenley - close by and one of the ring of RAF fighter stations defending the capital. The second was the Guards Depot, which had mushroomed into what was virtually a military town where thousands of recruits of the five Guards foot regiments were undergoing training. Both were specific, priority targets for low-level attacks.

Life in the Depot during these dangerous days is described by former Grenadier Fred Whittle who now lives near Kendal in Cumbria.

"The Depot had consisted of large, three-storey grey stone barrack blocks beside the parade ground, but when I arrived in July 1940 as conscript Grenadier recruit 2619944

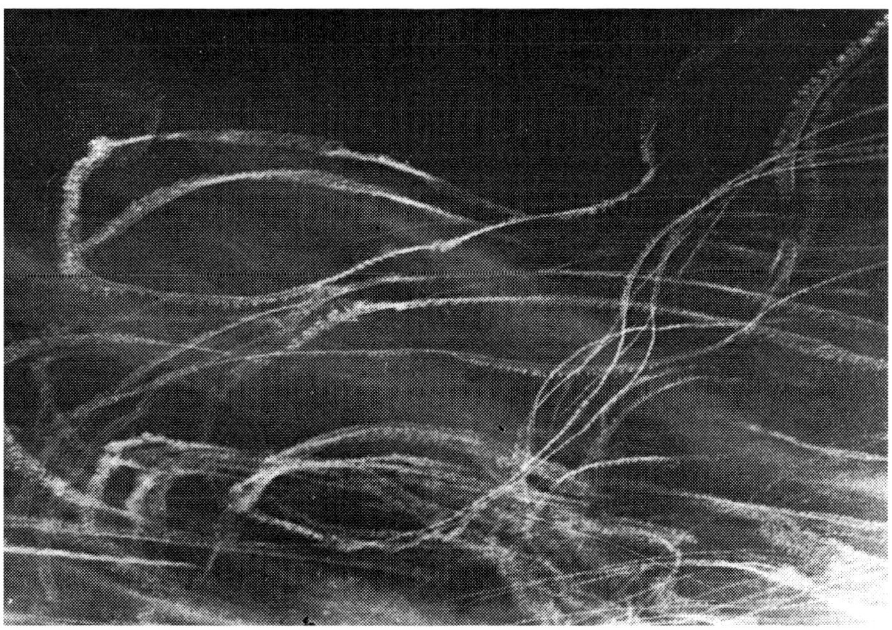

The skies over Caterham traced the deadly dog-fights of the Battle of Britain. Often the German bombers targeted the Depot on their way to the fighter stations at Kenley and Biggin Hill.

Whittle, F. they were surrounded by acres of Nissen-type huts, erected hurriedly to accommodate the huge call-up influx.

The intensity of the German raids increased through that Summer and the frequency of attacks severely affected training. Warning sirens meant we had to break off and take to the slit trench open shelters around the cricket pitch. Most raids were of short duration, but each cost at least one training session.

At night, we had to dress hurriedly when the alarm sounded, grab our steel helmets, respirators and rifles and take to the shelters. One night at the height of the Battle of Britain we had to run to the shelters nine times and all sleep was lost. After one 'all clear', I made my way sleepily back to bed on the top floor of one of the barrack blocks. In the dark, I started to climb in — only to find it already occupied by a very angry recruit. In the blackout, I had found the correct bed position, room and floor, but in the wrong block. I muttered a very quick apology and got out even faster, thankful that in the dark I could not be identified.

During one of the daylight raids, a lone German fighter swooped down and machine-gunned a crowd of us as we were running for shelter. Nobody was hit, but one of the barrack blocks bears the bullet scars to this day.

The possibility of invasion by enemy parachute troops was a frequent threat, and I spent one day in an orchard with sandbagged positions in case the threat became reality. No parachutists arrived, but we ate a lot of apples! Another night of raids was spent in and out of a requisitioned house in Coulsdon, guarding a Heinkel bomber which had crashed on an estate there, but without doing injury to anyone as far as I could tell. On other nights, we were less fortunate. One raid succeeded in destroying one of the smaller barrack blocks and a number of Welsh Guards recruits were killed or injured. The Depot chaplain kept a log of enemy bombs which fell within the barracks boundary. There were more than forty by the time I left the Depot.

Because RAF Kenley was only a few fields away, we saw the frequent 'scrambles' by our fighter squadrons each day as they went up to intercept the raiders. We also saw their victory rolls when they returned. But too often, it was noticeable that fewer came back.

Rumours of imminent invasion were rife. One such claimed a German fleet of troop barges had approached the English coast, but had been engulfed in burning oil on the surface of the sea. History is vague on this, but it is certain that Britain was in the gravest peril at that time.

The Brigade Squads at the Depot comprised the future prospective officers of all the five Guards regiments. They were youngsters like the rest of us and underwent the same hard training programme. Most, I expect, were former public schoolboys, but they shared our bawdy humour, expressed in the raucous songs of the time. More than once, we were in an open shelter next to one occupied by a Brigade Squad, and I heard for the first time, the words of the ribald ditty — sung to the tune of Colonel Bogey — about Hitler, Goering, Himmler and Goebbels.

Depot training was hard and thorough, and discipline would have seemed brutal to many civilians. But one experience showed me unexpected flexibility and humanity in the regime. A spell in hospital broke my training and my father, an ex-soldier, expressed his concern in a letter to my CO and asked if he could see me. It was almost unheard of for recruits to be allowed out of the Depot till they finished their training. But to my astonishment and delight, my Company Commander granted me a walking-out pass for three consecutive evenings so that I could meet my father in Caterham. Of course, I got this amazing concession only after I had promised to uphold the tradition of the regiment at all times while outside.

I believe I kept that promise."

Marking Time: Drill Sergeant B.E.Hillier, DCM, Welsh Guards, with pace stick, puts NCOs through a drill course. The hated puttees went out early in WW2 and were replaced by the more practical gaiters.

In the 1800s . . .

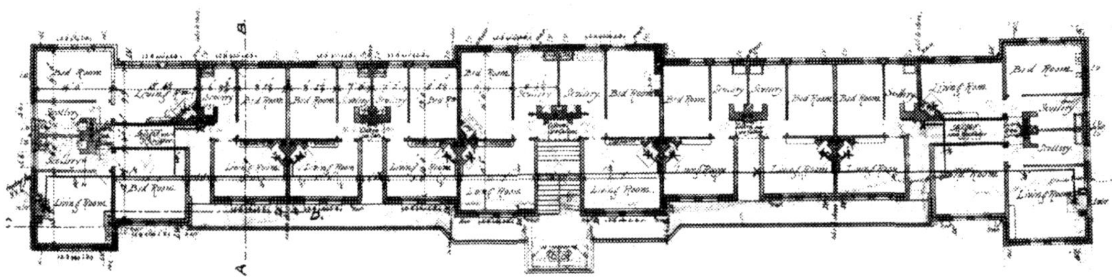

PLAN OF UPPER FLOOR.

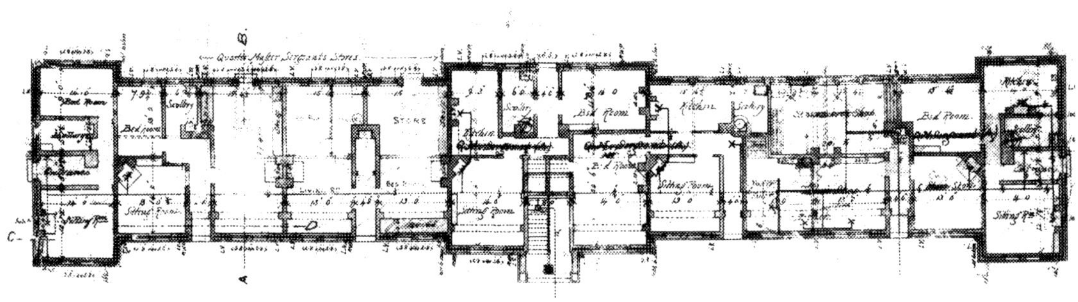

PLAN OF GROUND FLOOR.

From the Architect's (Major Seddon) drawings of 1875

The Caterham Boys

Boys' Service has been a tradition in the Guards regiments for many years. Youngsters who join are put in their own squads at the Depot and undergo training, separate from other recruits and on a programme designed to suit their immaturity and still-developing physique. Included in the programme is further education aimed at ensuring adequate standards in that area. Their instructors are chosen with particular care and their methods and progress are continuously monitored. Captain LE (Les) Burrell QM MBE was a Grenadier instructor in charge of a boys' squad at Caterham in early 1939, under the gathering shadows of WW2.

"Recruits were arriving at the Depot in greater numbers and I started a boys' squad in Woodford Block which was one of the Scots Guards' barracks. The squad consisted of recruits to all five regiments. Training boys at the Depot was an innovation — prior to that they had been allocated directly to battalions on enlistment. The Depot was being expanded rapidly and my squad soon found itself moved to a marquee which had been erected in the area between the sports ground and the public road.

Threat of impending war caused frenzied activity to cope with the huge expansion under way. Because the Depot would almost certainly be a target for attack, especially from the air, slit trenches were being dug in most areas. In the process of digging a trench system around the cricket ground, a network of pipes was discovered. This caused some puzzlement until the garrison engineer identified them as having served the hutted camp which had been built on the cricket field in WW1.

Large white marquees mushroomed on the cricket field to accommodate the vastly increased flow of recruits. A few days before the outbreak of WW2 the squad I was then training was moved to makeshift accommodation in the drill shed next to the sergeants' mess. It had been partially enclosed to provide some shelter. At the same time, an attempt to reduce the vulnerability of the large white marquees was made by painting them in camouflage colours. Bell tents were erected in the wooded area next to the rugby field. These were surrounded by a low wall of sandbags to give the occupants some measure of protection from the expected air raids.

I don't think these tents were occupied, certainly not before the Grenadier and Scots Guards moved to Chelsea Barracks in London. I believe these regiments returned to Caterham in 1941 when the Fox Lines camp was completed.

The Irish Guards' Roberts Block was damaged by a bomb one Sunday afternoon during one of the many raids on the RAF's fighter station at nearby Kenley. Later, during the blitz, Victoria Block sustained bomb damage and a bomb on the shelters in the cricket ground area caused a number of casualties. There were many raids on the Depot while I was there and before my battalion left to join 201 Guards Brigade in the Middle East.

George Duffin, now living in Tasmania, joined the Grenadiers at Caterham in September, 1945. He travelled 21 miles to Nottingham from his village, Edwinstowe, then by train to London and another to Caterham. It was the longest journey of his young life. He was 5ft 6ins tall, weighed just over eight stone, and he was the first boy to join his regiment after WW2.

"I remember the good times and bad times, but I look back to these days without any feeling of resentment. I think it was all character-building. We are better men for the discipline instilled at the Depot. I have never met a Guardsman who regretted his days at the old Guards' Depot, Caterham."

PROGRAMME FOR BOYS

WEEKLY PROGRAMME

Morning Periods	0845–0925 hrs. 0930–1015 hrs. 1045–1145 hrs. 1200–1230 hrs.	4 periods
Afternoon Periods (Monday, Tuesday, Thursday and Friday)	1345–1430 hrs. 1435–1515 hrs. 1520–1615 hrs.	3 periods
Wednesday p.m.	Organized recreation.	

Thirty-six periods in week to be made up as follows :

Drill	10 periods
Education	10 periods
P.T.	3 periods
Side-Drum / Flute	3 periods
Bugle practice	3 periods
Musical theory	2 periods
Baths	2 periods
Kit inspection	1 period
Religious instruction ...	1 period
Pay	1 period
	36 periods

N.B.—This is in addition to :

Evening Musical Practice on Monday to Friday when possible	1715–1800 hrs.
and	
Shining Parade ... ... (Monday to Friday when possible).	1800–1900 hrs.

Sergt. A. J. Astley's GUARDS DEPOT Boys Squad, Nov. 1945.

Boy. H. Scruton, P. Dyson-Holland, F. Boardman, A. Wintour, F. Tofts, G. Dullin, B. O'Hanlon
Boy. J. Crawford, D. Ford, A. Barrett, H. Calvert, A. Taylor, B. Mountford, R. Barrett, J. Moppett, F. Clare, P. Baker
Boy. J. Brookes, M. Mannel, Sgt. E. Waight (Supt. Sergt.), SGT. A. J. ASTLEY (Squad Instr.), Td.S. T. Rogers, Boy. R. Hillier, G. Britton

Corpl. J. Benson's GUARDS DEPOT Boys Squad, Dec. 1945.

Boy C. Pike, G. Burrows, J. Walsh, C. Smith, R. Purkiss, R. Cox, D. Vassar
Boy R. Bateman, C. Smith, B. Rockett, R. Jones, R. Robinson, D. Hethington, H. Westwood, H. Tansey, M. Thornton, J. Squire
Boy G. King, H. Wells, Sgt. E. Waight, CPL. J. BENSON (Instr.), Td.S. M. McGrail, Boy D. Horn, P. Taylor

In the 1800s

One recruit devised this 'Tommy Atkins' greetings card for the Depot.

Delivered from Disaster

The Luftwaffe raids on the Caterham Depot during WW2 caused much damage and tragic loss of life. However, although it demolished a married quarters building, the main impact of one bomb was the embarrassment it caused the Guards family living there.

The story of a very personal high explosive predicament is told by John Tipton whose late father, Welsh Guardsman George (Harry) Tipton, was a recruit at the Depot in the early 1940s.

Long after the war, his father was attending a St. David's Day parade when he was approached by a much younger man who said he remembered him from Caterham. This his father doubted, because the young man could not have been old enough to have served during the war years.

'No', said the young man, 'I joined the Welsh Guards much later as 61 Recruit David Jones, but I was there as a youngster of 15 or so because my father was 0034 CSM T. Jones, MBE, at Caterham in the early '40s.

The air raids were frequent then, and sometimes we had very little warning before the bombs started falling.

'As it happened, my mother was in an advanced state of pregnancy at the time and we were naturally concerned about her condition, and having to rush her to the shelters when the sirens sounded. Well, one night the bombs beat the sirens and one of them scored a hit on our married quarters. The explosion was deafening and when the smoke and dust cleared, my mother found herself sitting naked in the bath surrounded by rubble — and the recruits who came in to rescue her. This they did, and earned her embarrassed gratitude. One of the rescuing recruits was George Tipton'.

The story was rounded off even more happily. Not only did the lady survive unscathed, but years later at that St. David's Day parade, David Jones introduced George Tipton to the second person whose luck held when the bomb fell — his younger brother who was born shortly afterwards.

In the 1800s

PARCELS,
CATERHAM,
PER S.E.R.

GUARDS' DEPÔT,
CATERHAM,
SURREY

Bending & Stretching the Knees

Three Months in '44.

By a WW2 Conscript

I was called up and ordered to report to the Guards Depot, Caterham, in 1944. I was 18 and, apart from school holidays, I had never before been away for any length of time from my parents' home in Glasgow. Now, like it or lump it, I was about to become 2703196 Recruit Melvin, RGS, Scots Guards.

We were to be there at 12 noon on the specified day, and with our cardboard suitcases, it wasn't difficult for us to identify fellow conscripts on the train on our way down from London. The war against Germany was by no means over. On the Continent, fighting was about to reach a decisive stage and the South of England was cluttered in its byways and meadows with the huge allied build-up of reinforcement for the Overlord invasion of Germany's Fortress Europe. Beyond that, there was the small matter of the problem posed by the Japanese in the Far East, with the horrific catalogue of atrocity and inhuman treatment of all who opposed them.

We had no idea what to expect at Caterham and our nervy ignorance made some of us talkative, others withdrawn. All we knew was that we were joining one of the Guards regiments at the Depot, and our future beyond that was encompassed by the stark fact that we would be at the disposal of the War Office 'for the Duration of the Emergency'.

As we emerged uncertainly from Caterham Station, we were confronted by a sergeant in battledress who bristled with loud authority. He ushered us with little ceremony and evident disdain, into a small convoy of three-ton Bedford trucks which then made the low-gear ascent to Caterham on the Hill and the Guards Depot.

I was in the last truck, and I remember seeing the Depot gates close behind us. It was almost three months before I saw the outside, civilian world again. At that time, all five regiments of the Guards Brigade — Grenadiers, Coldstream, Scots, Irish and Welsh — were represented at Caterham. Thus, there were several thousand men undergoing basic Guards training at the Depot in 1944. It was hugely and, to us as new arrivals, chaotically congested.

In every space among the original stone-built buildings of the peacetime Army were crude wooden huts which had sprung up like an ugly undergrowth. Each was home to a squad of recruits and, because we were conscripts and the Depot authorities had decided to start us as they meant us to go on, there were no vestiges of pampered civilian life, such as beds, sheets and warm water ablutions. There was no gentle breaking-in.

As soon as we had been allocated a bedspace, we were taken to the barracks barber where we left most of our hair on his floor. Cap size was deferred until this visit had been made. Then we lined up for our jabs against just about everything the medical profession could think of, it seemed.

'Bed' was three planks on two three-inch trestles, three square horsehair 'biscuits' as a mattress and two blankets — one below, one above — and the pillow, a coarse linen bag filled with straw. A wall locker above each bed contained our personal belongings. There was no place for civilian clothes which had to be returned immediately to our home addresses.

We learned to maintain the creases in our battledress trousers by filling our mouths with water and running the creases through our lips. Required creases in blouses were dampened by shaving brush, and the whole uniform was then laid on our bedding 'biscuits' and covered by the under-blanket. Inexplicably, rheumatism was unknown.

From the Quartermaster, we each drew two battledress uniforms, denim trousers and blouses for hard wear training, shirts, socks, underwear, boots, a shapeless khaki beret cap to go with the denims, and the genuine Guards cap with distinctive peak. One cap badge, two identity discs with name and religious persuasion (no nonsense, agnostics had to choose one of the approved churches).

We were also issued with our AB 64. Part 1 was our formal identification and record of enlistment 'for the Duration of the Emergency'. Part 2 was our pay book. Part 1 made it clear that the duration of our Guards service was going to be the length of a piece of string. But one faint, distant hope of a return to civvy street was kept alive by the entry: Demobilisation Group 64.

Three things every Guardsman had to have on his person at all times; his AB 64, emergency wound dressing, and identity discs. Being caught without these could get you into the record books for punishment drills.

Our first two parades took place, not on the drill square but inside our hut. And, looking back now, their subjects served to bracket the extremes of Guards life. First, we took the oath of allegiance to King and Country. Then we, as individuals, were allocated our 'swabbing jobs'.

Administration of the oath impressed on us that we had joined a regiment whose history went back to 1642 and King Charles I. Furthermore, our regiment was part of the Guards Brigade, which comprised the Monarch's Household Troops. Thus, we were not merely called on, like the Line regiments, to defend King and Country, we had this unique responsibility to the Royals — hence the Brigade's monopoly on guarding Buckingham Palace.

At the other end of the scale from these fine thoughts, the swabbing jobs represented the housekeeping tasks we each had to carry out daily in the hut. Main features were keeping floor and windows proof against the snap inspections which were sprung without warning, maintaining the stirrup pumps and fire buckets of water and sand in total readiness, and furbishing and fuelling the two large black stoves which provided our heating. The ablutions had no such latter luxury. We learned to wash and shave in cold water ... carefully but, above all, quickly.

On parade inspection, those whose morning shave was deemed unacceptable were either 'in the book' or in line for extra drills or, in marginal cases, warned: 'take a pace nearer to your razor next time, lad, and take the safety catch off!"

The following weeks were a blur of parades, marching, weapons drill, interspersed with lectures on enemy arms and techniques, and our own regimental history, complete with battle honours. When it rained, we marched on. When it lashed down, we did weapons drill in the drill sheds, just off the main barracks square. The only way to get through it was to forget the outside world in a numbing concentration on life within a very narrow horizon, putting all other considerations on total hold.

Our unquestioned and absolute squad master was a corporal, a blue-jawed escapee from Glasgow's Gorbals, who sprang immaculate, each reveille to question our parentage and doubt our manhood. He chased us through each day, matching our 120-paces-a-minute on parade drills and stripping, cleaning and re-assembling British and German weaponry in speed demonstrations we couldn't equal. We never discovered how he came to be a full corporal (a rank attainable in the Guards only by being 'busted down' from sergeant).

His voice was hoarse bark, and at night we heard his fits of uncontrollable coughing. I learned later that his slum childhood had given him the tuberculosis which eventually killed him. It's a gung-ho cliché, of course — that men trained and led by unsparing and inspiring example will march on Hell. With him, I think we would have had a shot at it.

Between us and the Corporal was the Trained Soldier. His actual rank was Guardsman, the equivalent of Private in a line regiment. But he was in absolute charge of our hut. When we wanted permission to speak, or go to the NAAFI in the evening, or receive mail when he distributed it, we had to stand to attention and address him as Trained Soldier. In the evenings he was in charge of the 'shining parade' when we sat on the ends of our beds, blancoing our webbing and polishing our brasses. While we did, he marched up and down the hut rasping out questions about regimental history. The consequences of not knowing the answers made us learn fast.

Lt. Colonel Drury MC commanded the Depot. But he was a distant figure we saw once a week when he took Commandant's Parade with all five Brigade regiments, and the huge Joliffe Square (now long gone) echoed to the thunder of marching feet and shouted orders. Depot RSM D. Hobbs of the Grenadiers took the parades. Like all Guards RSMs, he was a fearsome man.

The British Army's senior warrant officer was RSM Ronald Brittain MBE, of the Coldstream Guards. He had been at the Depot in 1935 and was said to have the loudest voice in the Army. His reputation lived on nine years later.

Discipline dominated every aspect of our lives. The chances of being put on a charge, or 'losing your name', were legion and often very, very difficult to avoid. Going to the NAAFI one evening, with the Trained Soldier's permission, I thought I was stepping out briskly enough. A prowling CSM thought not and I 'lost my name' for being idle. No one was ever off-duty at the Depot. So, two extra drills. RSM Brittain once charged a Guardsman for being idle while cycling through the barracks. So much for free-wheeling.

The regular kit inspections were terrifying. They posed a serious risk of 'losing your name' and having to do punishment drills, alone or sometimes with fellow-defaulters on the main square. 'Getting a chasing' with rifle at the slope, all webbing equipment plus gas mask, at a marching tempo well above the standard 120 paces a minute, wasn't funny. Sometimes, men returning from defaulters' parade had to lie on their beds for half-an-hour before they could summon the strength to take their equipment off.

At kit inspections, everything had to be laid out with geometric and exact precision, otherwise the inspecting officer would flick the lot onto the floor with his swagger stick with the dreaded words, 'Take his name, Corporal'. All this may seem pointless nonsense, but under the impact of this regime, we achieved the highest standards of personal cleanliness and every item of our equipment was fully serviceable at all times.

Likewise, rank-and-file unofficial justice and punishment were swift. A soldier who attacked another in our hut was taken out and thrown into one of the huge emergency water tanks provided to meet the threat of the ever-present fire hazard among so many wooden buildings in close proximity to each other. He had been a member of a Glasgow razor gang and fancied himself a 'hard man'. Hut justice was quick, severe and sometimes resulted in hospitalisation. But I never heard of anyone breaking his silence in the subsequent inquiries.

Men living in close proximity, depending on each other 'not to drop the squad in it', in a regime of hard, round-the-clock discipline, formed a loyalty relationship which is difficult for outsiders to understand, and thus often criticised. To-day's psycho-babblers call it 'bonding'. It used to be *esprit de corps* before that concept became largely unfashionable among many civilian commentators.

Whatever, it was real for us and we came to understand the part it had played in the long history of the Guards Brigade, and the winning of forty-four Victoria Crosses.

Our hut was, effectively, our world. We were an assorted, random collection of reluctant recruits (we were not yet entitled to call ourselves Guardsmen) from various parts of the UK. There were several Englishmen, but most were Scots. The Sassenachs had basic problems of communication with the majority from north of the Border — but then I, a lowland Scot, found it difficult to understand hut-mates from the Western Isles and Aberdeenshire. I learned that accent is one thing, dialect quite another.

I had been a junior reporter on a Glasgow newspaper. My squad colleagues included an Aberdeenshire ploughman, a Dundee jute worker, an armature-winder from Stirlingshire, a nurseryman from Dumfries, a Jarrow steelworker, an apprentice tailor from Leeds, a pressure salesman from Barrowland — Glasgow's version of Petticoat Lane — a law clerk from Edinburgh, and a Lanarkshire boxer whose claim was that he had once fought on the same bill as Jackie Paterson. Other than our current situation, we had little in common. But the intensity of our training made us respect and value each other. We could argue and fall out with each other, but when it came to the outside world there was no question about it, it was us and them.

Our hut competed against other Scots Guards huts and other Brigade regiments, in various sports. Week-end athletic meetings in the Depot were pan-regimental and we ran, jumped and fought against all-comers. One famous victory for our hut, I remember, was in boxing (in the Depot, it was always heavyweight). Against all the odds, our Dundonian eliminated the reigning champion with an uppercut which grew in ferocity and power with each subsequent telling. Frankly, we were astonished, but our victor's nonchalant acceptance of congratulations was Oscar-winning stuff.

Our regular sessions in the gym achieved two results; we discovered unsuspected reserves of stamina and, despite a single-menu option in the mess hall we all put on weight, none of it fat. Our squad had a PE instructor whose broken nose and forehead scar tissue marked him down as a career boxer, but we all understood the penalties for 'silent insubordination' by this time and we gave him little trouble. In return, he sweated with us through our physical build-up. By the time he finished with us, we would have fancied our chances arrogantly against just about any opposition. He had fulfilled his task.

Entertainment opportunities at Caterham were limited in the main to occasional visits to the NAAFI after evening shining parade. But at the end of a long day of square-bashing, weapons drill, and working in the gym on parallel bars, climbing ropes, wall bars and horse, it didn't seem like privation. Most recruits were content just to lie briefly on their beds and read a magazine or write letters till the sergeant-in-waiting appeared thirty minutes before lights out to order everyone to 'stand by your beds!' for the nightly head count.

It was on one such evening that I first heard several side-drummers practising in a nearby hut. At first, it was just the kind of drumming which often accompanied our one-hundred-and-twenty marching paces a minute on the drill square. Then one drummer played a brief solo of rolls and paradiddles. One by one, the others followed, repeating the pattern of the first solo then improvising and embellishing it so that it became a kind of tuneless music. It developed into a contest lasting twenty minutes or so, and I realised for the first time just how skilled and ingenious the members of the drum corps were.

Like the rest of the home county population, the Depot had to take its chances with the first of Hitler's 'vengeance weapons', the flying bombs. Their track up 'bomb alley' towards London took them over Caterham and several fell in the depot, killing a number of Guards personnel. The stuttering sound of their engines, accompanied at night by their orange tail flames, were familiar aspects of life in 1944. When the engine noise stopped, everyone took cover, knowing the bomb was about to fall. Services for the casualties were held in the chapel, beside the main gate.

During my time at the Depot, no V2s — the second German 'vengeance' weapon — fell in the barracks. They had a longer range, and being supersonic, they arrived in London without warning. One of the first caused appalling casualties when it landed during a service in the Guards' Chapel in Wellington Barracks.

Incendiary bombs were frequent hazards in the wartime Depot, especially vulnerable because of the large number of wooden buildings. Fire-fighting drills were rigorous and regular.

We hadn't yet been judged fit to be seen by the outside world, but towards the end of our three months in the Depot we were allowed one small break in our isolation. After dark one evening, our Trained Soldier led us silently through the lines to a point in the wire fence. There he revealed a hole through which we could squeeze, one by one. The 'outside world' we found ourselves in were the grounds of The Fox pub on Coulsdon Common. We entered this establishment furtively and with repeated warnings from our Trained Soldier to keep quiet and not draw attention to ourselves. He made it clear that it would go hard for him if this act of Christian generosity on his part was discovered by the authorities.

Naturally, we were greatly impressed by his willingness to accept such risks for our sake, and we showed our appreciation by not allowing him to pay for a drink throughout the evening. We helped him back to the hut amid considerable noise-making which somehow failed to arouse any interest from the 'authorities'. Much later, the penny dropped. It seemed the Trained Soldiers operated a rota in order to avoid embarrassing squad 'overlaps' at the pub.

At the end of our basic training, two significant events occurred. We were instructed to take our caps with their straight-out-in-front peaks to the regimental tailor so that he could convert them to the 'cheesecutters' for which the Guards are famed. The peaks no longer jutted — they came straight down across our noses, so that the only way to see was to hold our heads up. We recruits all lusted after this visible evidence that we had 'graduated'.

The other event was the fact that we became entitled to the rank of Guardsman. Nevertheless, we were still constantly reminded of our fledgling state; men with even a few more months' service would advise us to 'wait till your number dries' in any situation requiring a put-down. But everyone with regimental numbers beginning with 270 deferred to the 269 men — all of them were Depot instructors, many with active and foreign service as regulars, before and at the outbreak of the war. A number had been at Dunkirk — men whose units had come back in good order and with all their equipment. We felt raw and inadequate beside them.

Came the day when we were allowed to apply for passes to leave barracks for an evening — half a day at week-ends. After our enforced confinement, being able to walk through the main gate had an almost cxotic quality. The Caterham pubs gave our first taste of freedom. After that, almost everyone scrambled onto the three-tonners provided as the SR station shuttle, and caught the first train up to London. In 'The Smoke' (the Capital's nickname before the days of central heating), our accumulated savings melted rapidly — the West End cinemas were beyond our pockets and bar prices seemed pitched at the American servicemen who thronged the capital. What we saw as this unfair competition got a few of us into trouble with the MPs but most of us managed to catch the last train back to Caterham.

There was no return transport to take us back

to the Depot and we learned to suffer the long, seemingly unending climb up Church Hill after an evening in London taverns. The blackout was total. There were no street lights and on cloudy, moonless nights the darkness was stygian — 'as black as the Earl o' Hell's weskit' was how one Mick sergeant put it. Navigation was problematical for some who had overdone it a bit, but it would have seemed a long climb, even if we'd spent our evening with the Church Army.

Guardsman Melvin

On our first outing, a Trained Soldier taught us the best way to make the climb. We linked up like circus elephants, holding the belt of the man in front and started off up the hill, unable to see anything after the station's dim lighting. More often than not on these occasions, the lead Guardsman kept to the road, but with anything between fifty and a hundred men in a variety of celebratory conditions, this linked-up 'crocodile' usually seemed more and more like an invisible and erratic conga. Even so, the whole operation was always completed in near silence, apart from a muffled expletive when someone kicked the heels of the man in front. Noise would have made it difficult for the guardroom duty NCOs to ignore our slightly dishevelled re-appearance at the main gate. Before turning in, we usually had to blanco away the heavy finger marks on the back of our web belts.

These interludes were the safety valves which maintained good order and helped us to accept the Depot discipline. The structured and relentless pace of our training programme occupied all the rest of our waking hours. It seemed to stretch ahead of us to infinity.

But the pressure and demands of the war on the Continent were continuous and increasing, and the day finally came when we were told to pack all equipment for the move to Pirbright, on the other side of Surrey. There we would undergo a more concentrated period of training and physical build-up — forced marches, regular visits to the assault course, close quarter fighting with grenade and bayonet, house-by-house clearances, and marksmanship training on the adjoining Bisley ranges. Our overseas posting would follow.

Suddenly, the Caterham Depot and the things which had seemed strange and intimidating to us — the old stone buildings and draughty, makeshift wooden huts, the vast parade squares and drill sheds, the gymnasium, redolent of disinfectant, rubbing alcohol and sweat — suddenly became the familiar and understood world.

Another world of imponderables and uncertainties lay ahead.

Postscript

I don't suppose the experience and impressions of this conscript Guardsman were very different in their essentials from those of any of the thousands of men who became soldiers at the Guards Depot, Caterham. The vast majority would have been regulars, signed-on for the period of their enlistment mainly seven years with the Colours and five in the reserve.

One major difference during wartime was the absence of the public ceremonial for which the

Guards regiments are famed. The standards of turnout in khaki battledress never faltered but the peacetime dress tunics and bearskins were in store for the duration in the Tower of London.

I did two 'Buck House' guards, one Trooping the Colour, and several Bank of England picquets at war's end while waiting for the magic 64 of my demob number to come up. But these were khaki ceremonies, lacking the spectacle of the real thing.

The peacetime Guardsman went through the same training process we did. Then he did his share of ceremonial duties here, or served a period in a foreign posting. Most recently, 'overseas' duties have meant Germany or Northern Ireland. Regulars and conscripts fought together for the last time on any scale in World War II.

The landlord of my local pub is an ex-Coldstream regular, but he began his boy's service at Pirbright. I tell him he missed something — a type of soldiering from another era. But he is proud of his service and above the bar he has placed a bearskin and red Coldstream tunic, beside the badges of the five regiments.

The pub's name goes back centuries — but I pull his leg and tell him they must have known he was coming when they called it The Bull. And when we argue, I'm sometimes tempted to advise this comparatively young ex-Guardsman to wait until his number dries . . .

Welsh Guards

Caterham Grew Loyalty and Pride

Welsh Guards Lieutenant Owen Griffith first saw Caterham when he was posted there as a result of severe wounds he sustained in Tunisia in WW2, during the British 1st Army's campaign against the German Army's elite Afrika Corps. Of the six Welsh Guards officers who had volunteered for North Africa in 1942, three had been wounded. The other three were killed. Judged to have no further active service role after seven months in hospital, he joined a Depot whose other officers had been recalled from civil life having served in WW1 two decades earlier, or were *hors de combat* because of recent wounds.

He found them all dedicated to the Guards Brigade and helping to run the Depot efficiently despite a huge overload in its ever-changing population of recruits under training.

"The Commandant was the Earl of Romney, a remote and formidable figure made all the more so by a black patch over one eye. Second-in-command was Sir Henry Aubrey-Fletcher, CVO DSO who had distinguished himself with the Guards Division in WW1. Invariably courteous and friendly, he had written several successful detective novels under the pen name, Henry Wade. Captain Goodhart-Rendell had been a prominent architect and a doctor of music, although he looked and dressed like a Prussian officer of the old school. There was also a Major Fryer who had at least one Military Cross and an outstanding record for bravery in the 1914-18 war.

I was attached to No. 8 Company, Welsh Guards, commanded by Major Harry Rice and with CSM 'Wacky' Jones in support. Both had been recalled from retirement and their fierce pride in the regiment and No. 8 Company was undimmed.

Two incidents showed me the humanity underlying the Depot's hard discipline - and the striking loyalty and pride in the Brigade it engendered.

One day, while I was taking a parade on the square with CSM Jones, one of the recruits suddenly sat down on the ground. I expected an explosion from the Sergeant Major. Instead, he said quietly: 'Leave this to me, Sir.' Then he asked the Recruit in gentle, almost avuncular tones: 'Boy, do you know what you're doing?' The Recruit replied, 'Yes, Sir'. The Sergeant Major repeated: 'Do you really understand what you are doing and what it means?' The Recruit replied that he did. The CSM's peremptory parade ground voice immediately returned and he marched the young man off under escort to the guard room. Afterwards, the Sergeant Major told me: 'it was clear to me he was a conscientious objector who had somehow slipped through the recruitment net. I wanted to make sure he realised the consequences of what he was doing'.

I was in the company office on another day when a young recruit in civilian clothes was marched in.

A few weeks of good Caterham diet, rigorous drill and demanding PT had built him up and his civvies were now too small for him. He stood there, rigidly to attention with tears streaming down his face. It had been discovered that he had been a coal miner in South Wales before he volunteered for the Guards. As mining was a reserved occupation essential to the war effort, the Regiment had no alternative under the law but to return him to the mines. He was heartbroken."

As he had no active service future in the Welsh Guards. Griffith was accepted by the Colonial Office to transfer to Uganda as a District Officer. He spent 19 years in the Colonial Service and a further 19 in the Diplomatic Service.

Today, Owen Griffith CBE MVO lives at Gerrards Cross, Buckinghamshire. Looking back he regrets missing his recruit training at Caterham.

'I was one of the last to go straight from university to Sandhurst and I had much leg pulling about that from my contemporaries who had been through the Depot's Brigade Squads. I know I would have been much better prepared for the rigours of Army life if I had done my stint at Caterham. It made a deep and lasting impression on all of us who served there.'

———oOo———

Guards' Humour

A squad instructor was to appear before the Company Commander for summary justice. The Sergeant-Major gave the command, 'March in!' The Sergeant-in-waiting, standing outside, having called the instructor to attention, shouted, 'Corporal Green ordered to attend, Sir!' On the word 'Sir', his false teeth shot out of his mouth in a graceful trajectory, hit the highly-polished floor and slid across, coming to a stop before the CO's desk. Not a facial muscle twitched. The CO rose from his chair, leaned over the desk to inspect the teeth and said, 'March those dentures out, Company Sergeant Major.' This the CSM did with a flick of his boot, and a very red-faced Sergeant-in-Waiting retrieved them, wiped them on his sleeve and popped them back in.

———oOo———

The Man Who Broke the Drill Pig's Heart

By The Editor

One Scots Guards conscript had the Caterham drill sergeants tearing their hair and caused a near-riot of laughter in the Depot ranks of the five regiments during WW2.

The first time this recruit formed up with his squad on the square, all seemed routine and normal. But when the squad corporal gave the order, Quick March, the centuries-old Guards pattern of precision drill came apart.

Like all his fellow recruits, one new soldier in the front rank stepped off smartly with his left foot — but, unbelievably, his left arm swung forward along with his left leg. And it went on: left arm, left leg. Right arm, right leg, on and on, and . . .

Until, that is, the incredulous squad corporal yelled the squad to a halt.

He then gave some fairly basic advice to the offending squaddie, and tried again. The result was exactly as it had been first time round. Several more attempts were made, but nothing changed. Except, that is, that the unorthodox marching technique disrupted the rest of the squad. Their own marching became ragged, the required as - one precision fell apart.

Higher authority was called in. A drill sergeant gave the offending recruit a short insight into what life would be like if he didn't conform. Still no change. So the luckless lad was extracted from the squad and received a 'chasing'— marching alone under the drill pig's orders at well above the 120-paces-a-minute norm. Even at speed, the left arm still went forward with the left leg, followed by the right arm with the right leg . . . Then the heavens opened and torrential rain began. The squad was dismissed, with dire warnings for the 'morrow if the offender still didn't shape up.

Back in the squad hut, reactions ranged from 'Just keep trying, lad, you'll get the hang of it', to 'You trying to work your ticket, eh?' One or two comics tried to mimic the outlandish marching technique. They found it impossible to keep it up for any length of time. The speculation and banter went on till lights out, its subject admitting his fault, but claiming he couldn't do anything about it.

The following day's performance was virtually a re-run. After twenty minutes of total frustration for the Company Sergeant Major, the recruit was taken back to the drill shed. There, he was given carefully choreographed slow-motion practice, no doubt designed to convince him that not only the Guards, but world armies going back to the Roman legions had got it right, and he hadn't.

The drill pig's day brightened. In near freeze-frame slow motion, the right arm went forward with the left leg, followed by the left arm and right leg. This breakthrough was noted by just about everyone else on the parade ground, and dozens of eyes swivelled to watch the one-man awkward squad repeat the successful, if otherwise unimpressive technique up and down the drill shed.

After a brief halt and some encouraging words from the now optimistic drill pig, the progress was put to the test at normal marching pace ... and it all immediately reverted to left arm, left leg; right arm, right leg. The drill pig persevered but to no avail — and by this time there were no NCOs or WOs inclined to volunteer to assist with the problem. The Guardsman's red face was more than matched by the pig's as his blood pressure soared. And on the square, discipline and good order came close to breakdown as dozens of jaws clamped down to stifle laughter.

Back in the squad hut that night, the unfortunate Guardsman (by then, his fellow recruits had concluded that his military dysfunction was genuine and involuntary) got a lot of sympathy. But there was still huge enjoyment of the drill pig's predicament.

Next day, the focus of everyone's attention disappeared. He was ordered to stay put while the squad went off to continue the daily training regime. When they returned, he had gone, along with all his kit and personal belongings. We never discovered where, although speculation had him in the glasshouse, in a special unit undergoing psychiatric treatment, even posted down the mines. Some even suggested unkindly that the plot was to allow him to fall into enemy hands in order to foul up Gerry's basic training programme. But the whole episode closed with none of us the wiser.

Sometimes, in dreams, I see the shadowy figure in the drill shed, still doing his ultra-slow march with total concentration and precision.

And I'm sure the drill pig does, too . .

In the 1800s . .

The Commandant's lady escaped with little worse than ruffled dignity, judging by a junior officer's light-hearted record of what happened when an axle broke while Mrs. Crabbe was setting off with her children for an outing in 1885.

Lèse-Majesty in the Mess Hall

The ravenous appetites he and his mates developed from recruit training at Caterham Depot in 1940 are recalled by 2619121 ex-Grenadier Guardsman J. Clarke. So are the chants of WHY ARE WE WAITING? when the food was slow to arrive on the messroom tables. He particularly remembers one of these days . . .

"We had been told during the morning that we would have a VIP visitor during our mid-day meal, but most of us had either forgotten or were too hungry to care when we sat down with no food in sight. After some time, and still no meal, the lads started chanting and banging their mugs and eating utensils on the tables.

Suddenly, there was total silence. All heads had swivelled to the messroom door and mouths, caught in mid-chant, hung open. Walking into the room was Her Majesty Queen Elizabeth, now Her Majesty The Queen Mother. You could have heard a pin drop. The Queen must have heard the unruly noise before she entered, but gave no sign of it. She smiled and chatted with a few of the lads before she waved to all of us and left.

The meal was then served quickly. Somehow, I think someone later got it in the neck because the food had not been on the tables when Her Majesty arrived.

There was also a day when I nearly got it in the neck! I was on a week's duty, helping to man a high-powered water pump which was part of the Depot's fire fighting defence against the frequent Luftwaffe attacks. We had to test the pump, of course, and that meant wheeling it out of its shed and throwing its intake pipe into one of the big emergency water supply tanks. Having started the pump engine, three Guardsmen then directed the powerful jet onto the 'fire' — in a practice situation, back into the tank.

Suddenly, the hose I was holding on to began to pull me all around the area, the water going everywhere except back into the tank. I realised that the Corporal in charge had increased the engine power, and the two other Guardsmen had let go. Before the pump could be stopped, I had sprayed one of the barrack rooms. Its windows were half-open but, fortunately for us, the occupants were not in. Believe me, we set records getting that pump back into its shed!

Another near-disaster was the night I went to the Palais-de-Dance in Croydon. I was talked into catching a train back to Purley instead of the last bus. A mate and I thought we were nearly there when someone told us the first stop was Redhill.

We were in trouble. There was no way we could get back to the Depot by midnight. In the blackout, we met a railwayman and explained our problem. He told us a train driver was getting his engine ready to do the morning paper delivery run. He might take us back to Purley. Fortunately, he did, but we were very, very late when we reported to the Depot guardroom.

The Sergeant was sitting behind the table. He asked: 'What leave are you on?' 'Midnight, Sergeant,' I replied. Then I told him of the night's events. 'Go to your bed, lad', he said. I was relieved. I left my cigarettes on the table for the Guard.

The Depot clock struck two, but it was a lovely night after all, and the moon came out as I reached the barrack room. Then, disaster. From the block stepped the tall figure of Captain 'Benny' Ford. He returned my salute and said: 'Goodnight, Guardsman — or should I say, good morning?' I replied: 'Good morning, Sir.' And that was all.

It was with real sadness that I heard Captain Ford had been killed at Anzio, when I was with the 6th Battalion at Mintero in Italy."

———————oOo———————

Guards' Humour

Someone hit on the idea of putting a small coin in otherwise empty rifle magazines to improve the 'dramatic' sound of heel-of-hand-hitting-rifle-butt in coming to the slope - a satisfying, unified clatter intended to bring joy to drill pigs, warrant officers and assorted commissioned ranks overseeing Caterham parades.
Sadly, not so.
We who tried it got extra drills, presumably for 'employing artificial agencies in breach of the sacred, time-honoured methods.' At the time, it was expressed a little differently - 'You dozy men; you sound like skeletons shagging on a corrugated iron roof!!'

———————oOo———————

Take my Mother-in-Law

The WW2 concerts by ENSA were a welcome distraction from the dangers of active service and the rigours of training. Many entertainers who were household names gave their time and talents; but so did a few who wouldn't even have had end-of-the-pier bookings in peacetime. Remember the concert party in 'It Ain't Half Hot, Mum'? David Wessel sent this recollection from New York . . .

"It was December, 1944, and I was 2725292 Recruit Wessel, FD, Irish Guards, in one of the Brigade Squads. It was a desperately cold winter and, therefore, there was more misery than in just trying to become a soldier, with all aspects of the new and exhausting life to contend with.

One evening — a weekend, I believe it was — we were told there was to be an ENSA concert in the Depot gym for our benefit. This sounded like a great idea and a welcome relief from the hard grind of barrack life. It was just before Christmas and we had all heard some wonderful ENSA productions broadcast by the BBC, so three of us from the squad went, full of high spirits and festive expectation.

The show started. It was bad, and it didn't get any better as it progressed. Bad became awful. Came the intermission and with it the prospect of relief, we thought. We had seen all we wanted to see of such poor quality 'entertainment', so we headed for the exit.

That was as far as we got. The burly NCOs of the Depot police were covering the exits and promptly returned us to our seats. We sat down again. Recruits don't argue with NCOs, let alone police NCOs. I suppose everyone knew how dreadful the show was and realised that if we were allowed to go, the rest of the audience would do likewise. The remainder of the evening we endured with gritted teeth, a truly 'captive audience'.

Looking back, it makes me laugh — something I certainly didn't do during the concert. It was one time when it wasn't 'all right on the night'!"

In the 1800s . . .

(not too successfully, from his expression)

It'll Make a Man of You, Lad!

The old recruiting sergeant's promise may have glossed over some of the less than wholly-enjoyable experiences between taking the King's or Queen's Shilling and attaining the desired state of masculine excellence, but it is indisputable that 99.9 per cent of men who joined at the Caterham Depot, left it much stronger, fitter and with confidence based on a vastly improved understanding of their own physical and mental resources.

Take the unlikely case of Dennis Haydock of Strensall, Yorkshire, who registered for military service on turning 18, and whose call-up papers instructed him to report to the Depot on July 2, 1942, when WW2 was at its height.

"I couldn't believe it. A colleague at work assured me that Caterham was the Guards' Depot but although I had passed my medical as A1, I was very thin and weighed only eight stones — not nearly robust enough, I thought, for a Guards' regiment.

At the time, the farthest I had been from my home in Sheffield was Scarborough, so a journey to London was exciting. So, still filled with doubts, I duly reported to Caterham barracks where I was directed to a gym and given three boards and two trestles — my bed for the next few nights! There were at least 200 of us in the gym, and I remember we were given a concert by the band of the Coldstream Guards. Then we were paraded, allocated to various regiments of the Brigade (mine being the Coldstream) and put into our squad huts.

From this point, we ran everywhere. We ran to collect our blankets, our mattresses, or 'biscuits', and to fill our pillows with straw. We ran to the barbers, the medical centre, the dentists, the equipment store and the armoury.

We stood still long enough to be 'introduced' to our Trained Soldier, our Squad Instructor and our Supervising Sergeant. Our run to the equipment store was rewarded by the issue of some second-hand webbing which had been blancoed a different colour to the one required, so we had to scrub it clean for a start. But it was a warm day and it dried quickly and was blancoed the correct colour.

Our rifles were new — and covered with thick grease which had to be cleaned off thoroughly. The wooden stock and butt then had to be burnished with dark tan boot polish. This item we got on our run to the NAAFI, along with black boot polish, dubbin, Bluebell metal polish, and a block of blanco. All of these necessities came out of our meagre pay.

We thought the barrack room was already spotless. We were wrong. Every morning, we had to swab out our own bedspace before breakfast. The fire buckets shone like silver, the stoves were polished black in a pure white surround and the handles of the shovel, mops and broom gleamed like prized furniture.

After breakfast came various aspects of training. We paraded on the square for drill, we ran to the gym for PT, and to another hut for weapon training, and ran back to our own hut, were dismissed, and immediately ordered out again for more PT. We became experts in speed-changing into PT kit, into uniform for drill, denims for weapon training. By this time, we felt we could run for Britain.

Every Saturday morning, a 'swank' parade was taken by the Commandant and we marched to the music of one of the Guards' bands. Often, we were watched by men from line regiments, or the RAF Regiment. First, we marched without rifles, then with rifles and, finally, in passing-out week, with rifles and fixed bayonets. On the square, we would feel the sergeant's hot breath behind us as he barked out questions like, 'What's the name of the regimental slow march? — What was the regiment's first battle honour? — When was the Coldstream regiment formed? — Who was General Monck? We learned fast. We had to.

Depot training came to an end and we prepared to leave for the Training Battalion at Pirbright. First, however, we were given another medical examination. To my amazement, I discovered that the weedy, eight-stone lad who had arrived three months earlier, was leaving within a few pounds of eleven stones — and there wasn't an ounce of fat in that!

I never saw a recruiting sergeant, and nobody had ever promised the Coldstream would make a man of me. But that's what happened. Our squad left the Depot in full service marching order and marched to the station, en route to Pirbright and, from there, to active service in Europe with the 1st Armoured Battalion.

I came back to Caterham, briefly, before I was demobbed in 1946, a year after the war ended. I saw the recruits arriving with each intake and watched them finding their feet. And although I never spoke the old recruiting sergeant's promise, the words always went through my mind ...

——————oOo——————

Guards' Humour

A Sergeant of the Guard at Caterham used to advise young Guardsmen 'walking out' for the evening that he expected them back at the due time and in reasonable condition. 'Like your rifles, lads - clean, bright and only slightly oiled!'

——————oOo——————

Tragedy and the RSM's Son

Raised in the old Depot, one of Michael E Johnson's early memories is of his Company Sergeant Major father weeping silent tears, two days after the Allied invasion to liberate Europe in 1944. News had just reached Caterham that the squad his father regarded as his best ever had gone ashore in the first assault and had been wiped out to a man in a machine gun trap. Thirteen years later, Michael was back at the Depot as 23430585 Recruit Johnson ME, Scots Guards — but only just. He wrote these recollections from his home in Gawler, South Australia.

"I had been the mascot of that ill-fated platoon. Then on D Day + 2, my father heard that thirty minutes after they waded ashore, they were all gone.

In the years which followed, my family travelled around, living in a number of places in the UK. When I returned to the Depot in 1957, my father had retired as a Regimental Sergeant Major in the Scots Guards. Why that regiment, I don't know because his father had fought in the Coldstream Guards in WW1. My older brother had just been drafted into the Grenadiers to do his National Service, but I decided to follow my father, and opted for his regiment.

Oh, the whirlwind of the first few days! Rushing here and rushing there. Papers to check to make sure we were who we said we were (as if anyone would be daft enough to come if he didn't have to!).

The Quartermaster's store — 'Keep in line. Hat size? You look like 6 7/8, that'll do. Boots? — 8 ½ — right, here's a pair of nines, your feet will swell soon enough!' Everything from greatcoat to drawers, cellular — 'sign here' and off we staggered clasping our kit, not knowing we had just signed up to our future torture.

The medical examination — 'Breathe in, hold it, cough, colour of eyes? Distinguishing marks? Scars, laddie — have you any scars? Open wide, say, ah! How tall? Stand there while we check — 5' 8 ½"! They've got you down for the Scots Guards, but their minimum is 5' 9". You'll have to go elsewhere.'

The Sergeant said it — I'm not tall enough. I would never have made the Grenadiers, anyway. Thank God I can escape; I can leave this bedlam. When can I get out?

'Excuse me, sir, this man is 5' 8½", can you sign his rejection, sir.'

'Right, Sergeant, let's have a look. Ah, I see you're a volunteer, lad. Why did you want to join the Scots Guards? Oh! I see, your father an RSM, you say. Just stand up against the scale again. Sergeant — be a bit more careful. See, this man is 5 feet 9 inches. He's in!'

Later: 'Right, lads, you've had your injections. Now, you're officially on light duties till they take effect, but there's no real need. In fact, what you need is a little light exercise to get the blood to race through the body. Rifles up above your heads! Now, running on the spot, begin! Knees up, knees right up to your chests — get those arms up, get those knees up. No, don't touch him — let him lie there — pick those feet up!'

Up before dawn and rush to the ablution blocks — wash, shave (and make sure it's close and no cuts!) — freezing cold, pouring with rain. Bash your tabs — permission to enter, Trained Soldier? No wonder Asian 'flu ran through the Depot. The epidemic caused chaos to the training schedules. I got it early and went into hospital. The rest of the squad came down with it, but the hospital was full by then. The rule said if you spent too long in the hospital or guardroom before the six-week inspection, you had to repeat three weeks' training.

I rejoined my squad from hospital, and got the 'flu again. While I was down again, the squad was back-squadded without me. In the end, I was back-squadded twice and spent 16 weeks in the Depot before I passed my six-weeks' inspection and was reckoned fit to be seen by the outside world.

Other memories: The pride of marching off the parade ground at that six-week mark, the smell of 'burning boots' and the gradual build-up of layers of polish till that gleaming shine which exceeded everything achieved before. The care of walking on the newly 'bumped' floor before the officer's inspection; the gleam of the chin strap in the light of the barrack room, sitting astride the bed on shining parade; the ice on the drill square — 'dig-in those heels and you won't slip! Am I hurting you? No? 'Well, I should be — I'm standing on your hair. GET IT CUT!'

And then, the final parade, with every other squad watching our final inspection. The swagger as we march back to barracks. Forget the drill sheds, forget the knees-up, double-time, rifle extended above your head, while the weaker collapsed around you. We had taken everything they could throw at us. We were Guardsmen, finally, ready to join the rest of the best in the battalion.

Those gates looked great as I went through them for the last time — a very different person to the one who had entered ...

Ringcraft and 'Gallantry'

It was absolutely standard for would-be recruits to arrive at the Caterham Depot with mixed emotions. Joe Harding was an exception. When he walked out of civilian life and reported to the main gate guardroom in early 1943, his mind was focused on just one thing. And that thing blotted out all but one emotion. He was, in his own words, 'a very scared young man'.

He wasn't afraid of the hard discipline and rigorous training he would undergo. Nor the fact that he would certainly be among the assault troops in the forthcoming invasion of the Continent. His only fear was that he would be found to have lied about his age in his application to join the Grenadiers.

Now living in London, Ontario, he recalls:

"I had used Milton disinfectant to remove my actual date of birth from the entry papers, after my parents had signed them. I was terrified that someone would discover what I'd done and it would be back to Civvy street. My worry wasn't lessened when I found that I was in a squad with four former police officers!'

The fear of discovery and rejection lessened as time passed, but Joe didn't feel completely safe for over a year and he achieved the minimum legal age of seventeen-and-a-half. One of the ex-policemen became a major influence in Joe's Depot life. He had been a police boxing champion, and took Joe under his wing when the youngster showed an interest in the ring. The rigour of Depot life allowed little free time, but much of it was put to good use.

'In the precious evening hours between shining parade — when we cleaned and polished our kit — and lights out, we dashed along to the gym. Despite the long, exhausting days we trained hard, night after night. I learned a lot.'

As proof of that, Joe Harding later became Battalion, and then Brigade heavyweight boxing champion.

The time came when his superiors judged him fit to be seen outside the Depot, wearing the Guards uniform. The occasion bruised his pride, and taught him a hard lesson about human nature. 'The first time I earned a walking-out pass, I went to Croydon. I was walking along the main street when suddenly I saw a man knock a lady to the pavement. I went to help her — and she attacked me, telling me in no uncertain terms to mind my own business!'

Joe swam for the Grenadiers' tug-o'-war team in the Depot pool. 'The teams consisted of either three or four men. Attached to the rope were loops which we slipped over our shoulders. And, of course, the teams never saw their opponents during the contest because they swam away from each other. It was the most strenuous sport I ever tackled; sometimes we had to be helped from the water.'

Two recruits in Joe's squad stand out in his memory. Albert Nightingale stood six-foot-nine and weighed eighteen stone. His boots were sixteens and his cap was eight-and-a-quarter inches. 'Also in our squad was a man who had the world's worst memory,' Joe recalls. 'He drove the Trained Soldier crazy because he could never answer any of the questions about the Regiment and the Brigade. It was so bad that each time the squad was up for inspection, in the fifth, tenth and twelfth weeks, he was sent on dental parade while the rest of the squad answered the officer's questions. But, came the final inspection in week fifteen and our absent-minded friend had to attend because there were no more dental parades.

'We were drilled under the gimlet eye of the Company Commander. Then the interrogation began. We were all praying that the officer would miss our problem squaddie, but our luck ran out. He was asked a series of simple questions but he wasn't able to answer any of them. In desperation, I suppose, our Company Commander asked him the name of our Trained Soldier.

His reply made every last one of us cringe. 'I don't know his real name, Sir, but we all call him Buckteeth."

———oOo———

Guards' Humour

It seemed a good idea at the time. One man in the middle of the squad would count out the time, sotto voce - 'One, toopthree one, toopthree one, toopthree one', so that the rifle drill would be absolutely spot-on and as totally co-ordinated as humanly possible. But among their other unloveable qualities, drill pigs have ears like Jodrell Bank. We got a real "chasing" on the square in which the standard 120-paces-a-minute went up to something like 250. 'You're a bunch of chancers,' roared the Pig. 'And, anyway, that trick made the French fall about at Waterloo. Do me a favour. Try to come up with a gimmick without whiskers!!'

———oOo———

The Demon Bowler and the CO

Aneurin Hopkins became 2739088 Welsh Guardsman Hopkins A at Caterham in 1943. He was 17 and 'straight from school in the heartland of Welsh Wales'.

He saw action with his regiment's 1st Battalion from Normandy to the Rhine in WW2 and eventually finished his service back at the Depot. Among his many memories of Caterham, two are of cricket. Both recall situations which looked like having serious consequences, but Aneurin looks back on them now with amusement.

"It was 1947, and it was to be a great day at the Depot. India had achieved independence, had her own Army and could train her own officers, so there was no need for the Indian OCTU. Huge marquees had been erected as part of the celebrations, a highlight of which was to be a cricket match. Army 'brass' and senior Guards officers mingled with civic chiefs and other notables in a relaxed atmosphere.

Norman, one of my mates, was to bowl for the Depot against our opponents. We batted first and, without troubling 9, 10 and 11, we made well over 200. This had given Norman ample time to sample the free hospitality of the drinks marquee, and by the time of the declaration he was in a very expansive mood. He walked unsteadily to the wicket and our captain, Major Romilly, handed him the ball before taking up his own position behind the stumps.

Overjoyed at the opportunity of exercising his demon bowling against an officer wicket-keeper and eleven officer-batsmen, Norman accelerated into his run. In his delivery stride, however, something went wrong between his brain and his legs. He landed flat on his face with the ball still in his grasp. Everyone shuddered. He tried again. This time he delivered the ball, but it narrowly missed the heads of batsman and wicket-keeper, landing first bounce about ten yards in front of me, right on the boundary. With the greatest difficulty, he managed to deliver six 'good' balls and was then banished to replace me at fine-leg on the boundary so that I could bowl.

With only a few halves of beer under my belt I took a wicket with the last ball of my first over. My team-mates congratulated me and Norman, in a very excited state, came in from his distant position in a staggering run. I thought he was going to add his congratulations but, horror of horrors, I realised he was making straight for the Depot Commandant, Lt-Colonel JC Windsor Lewis, DSO MC, resplendent in Etonian cap and cravat.

Before I could stop him, Norman's enthusiasm overflowed and he gave the Commandant a hearty thwack between the shoulder blades. The Colonel stumbled and Norman shouted, 'Isn't he a bloody good bowler, Cush!'

Time, stood still. So did everyone on the field — officers and other ranks. I think we all stopped breathing. I thought, 'I've seen everything now' and the consequences of Norman's act began to cross my mind in a hellish parade, like the visions of a drowning man. This was surely an offence, unprecedented in the history of all five Regiments of Foot Guards. In my state of shock, I saw Norman court-martialled, suffering years in a military prison, blindfolded before a firing squad, drummed out of the regiment in disgrace.

Suddenly, we all began breathing again. The Commandant was smiling! Then he called a couple of Guardsmen from the crowd and asked them to help Norman back to his barrack room, advising him as he went to 'sleep it off'.

Making my way back to fine-leg, still in a state of shock, I told the departing figure of the demon bowler: 'Norman, you're a lucky so-and-so. Only you could have got away with that! And you probably won't even remember any of it tomorrow!!'

*****

During the summers of 1946 and 1947, I played cricket for the Depot and the Household Brigade, and took quite a few wickets. As a result of that success, I was asked by the Commandant one weekend to bowl at him in the nets at four o'clock on the Saturday afternoon. I felt honoured but at the same time, I was cursing my luck. I would have to forget plans for 'walking out' with my mates. Still, I consoled myself with the thought that the Colonel was not a bad chap. He had a fine reputation in the Regiment from his wartime service and, if I bowled well at him I might be considered for a regular place on the Brigade team.

Waiting for four o'clock in an empty barrack room was tedious and I whiled away the time reading and I must have fallen asleep.

I awoke with a start, to find it was ten minutes to four. Hurriedly, I dressed in my cricketing whites and rushed to the nets just in time to see the Colonel drive away in his car. I was mortified. I knew the rules. You were 'late on parade' if you were not at the appointed place ten minutes before the parade began.

Full of apprehension, I waited in the empty barrack room. Then the Orderly Sergeant marched in, looking very serious. 'What the hell have you done?' said he. 'The Colonel wants you to report to his house at the double. Get yourself properly dressed!'

The door was opened by the still-smouldering Commandant who told me to report to the Guardroom to be placed under 'close arrest'. This I did, to the evident amusement of the Sergeant of the Guard. It was the first, and only time, I experienced the solitude and indignity of being a 'prisoner'. I spent the time wondering what charge I would face for disobeying an order given by the Commandant of the Depot? And what would follow? Extra parades? Extra drills? Pack drills?

I was marched, under escort, to Commandant's Orders. The Great Man sat relaxed and impassive as he regarded me. 'Have you anything to say, Hopkins?'

I kept my gaze riveted on the wall above his head. 'No, Sir.'

'I think you have had sufficient punishment. Let this be a lesson to you. Report back to duty!'

Regimental Sergeant Major Rooney roared the marching-out commands in his thick Irish brogue. 'Jesus and Mary!', he exclaimed when we achieved the veranda, 'you're a very lucky young man. You got off lightly because he wants you to play against the Greenjackets tomorrow!'

Back in the company lines I met Norman, the demon bowler. There was a twinkle in his eye, I think, when he wondered out loud why he hadn't been asked to give 'Cush' a net.

The next time I was on Commandant's Orders was when I got my discharge in October, 1947. He spoke very kindly to me, shook my hand, and wished me well.

In my discharge book, he had written: 'A diligent, intelligent soldier who worked hard as a fast bowler throughout a long and hard summer.'

It was perhaps the finest accolade he could give."

Many men from other regiments — officers, warrant officers and NCOs — attended Caterham training courses under Guards instructors. This WW2 group were from various Canadian regiments during the Allied build-up for the invasion of the Continent.
In the centre is the Earl of Romney who was Depot Commandant from 1928 to 1931 and from 1939 until 1943. He was the only officer to serve twice as Commandant. On his right, Drill Sergeant T.W. Kirby, Coldstream Guards; on his right RSM P. Dunne, Welsh Guards.

Major Derring-Do . . .

From the Croydon Chronicle, Sept 1890

———:o:———

Matters are quiet at the Depot The new Commandant, Major HH Colville, has fallen into the routine work very quickly. He has done some smart and plucky things in his time. He it was who decided after marriage, on spending his honeymoon in the clouds, and for this purpose proceeded to the Crystal Palace and, after the wedding breakfast was over, sped away heavenwards with his bride in a balloon. He is an experienced aeronaut and he accomplished his bridal trip without any mishap. It will be fresh in the minds of our readers that the gallant Major created some sensation by his successful attempt to cross the Channel in a small canoe.

———:o:———

Members of the Guards Depot Women's Club

JULY, 1943.

G. & H. BUNCE PHOTOGRAPHERS

Mrs. Phillips, Mrs. Newland, Mrs. Cook, Mrs. Frankum, Mrs. George, Mrs. Curzon, Mrs. Hallam, Mrs. Abbott, Mrs. Wood, Mrs. Hawksworth Mrs. Ginns
Mrs. Taylor, Mrs. Solly, Mrs. O'Gorman, Mrs. Smith, The Countess of Romney, Mrs. Villars, Mrs. Dalgleish, Mrs. Hardy, Mrs. Nott and Malcolm, Mrs. Astle
Mrs. Tooke, Thelma Villars, Billy Taylor, Joyce Webber, Mrs. Webber, Sister Schofield, Mrs. Thompson

Depot wives did much charitable and morale-building work, especially during wartime. Caterham's social life was an important factor in the blackout and rationing years, and the wives helped to organise functions and entertainment which were enjoyed by many townspeople as well as the garrison. This 1943 group were members of the Guards' Depot Women's Club, with some of the Depot children. The Commandant's lady, the Countess of Romney is fifth from the left, middle row.

In the 1800s

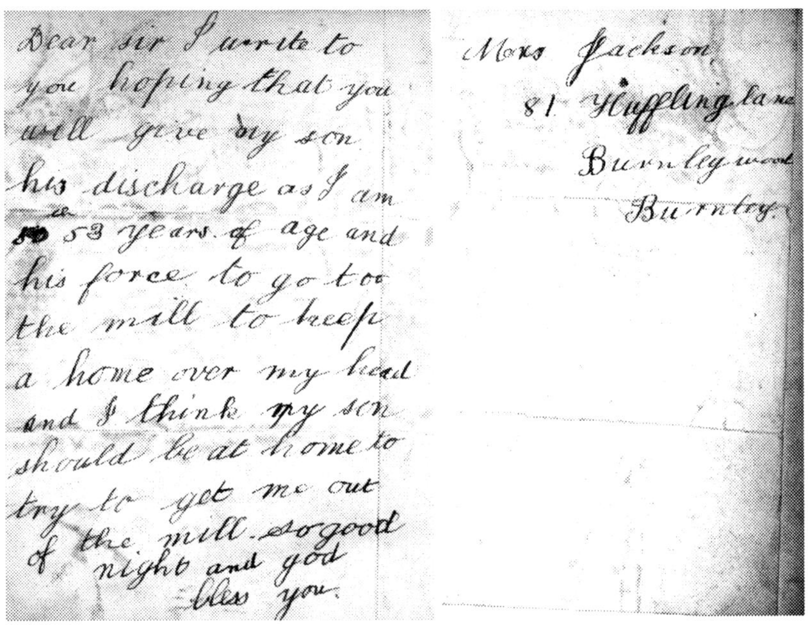

Dear sir I write to you hoping that you will give my son his discharge as I am 53 years of age and his force to go too the mill to keep a home over my head and I think my son should be at home to try to get me out of the mill so good night and god bless you.

Mrs Jackson,
81 Huffling lane
Burnley wood
Burnley.

When this widow wrote to the Commandant, fifty was an advanced age, compared to today's female life expectancy of almost eighty. That Guardsman Jackson got his discharge is unlikely.

Blackout Perils

Strict blackout laws were in force during WW2, when air raids were an almost nightly occurrence. But the unrelieved darkness created its own hazards and dangers for everyone outdoors after sunset. 2624778 ex-Grenadier Sergeant Sid Tidman remembers his first night outing from the Depot, and a frightening experience . . .

"Ten weeks squadded, and we were to be allowed out of the Depot for the first time! Two of us decided to go into Croydon for the evening, but we got off to a bad start. The Sergeant of the Guard turned us back twice because he judged we hadn't approached the guardroom smartly enough. But at last we were out, and caught a train at Caterham. We settled down in the darkened carriage and relaxed, looking forward to our first taste of freedom'.

Because of the constant danger of air attack, there was one very small blue light in the compartment — enough to show where the seats were, but little else. It hadn't been a good day for our squad. We had earned a 'chasing' on the square and we spent the time discussing it and airing our views about our instructor. A sympathetic voice from the far corner asked us who had inflicted this treatment, and one of us named the son-of-a-bitch, adding some more colourful descriptions, plus a short run-down on what he would like to do to our tormentor.

When the train stopped at Croydon, another small light came on to let passengers get off safely — and we got a shock. In the corner sat a burly Guards sergeant, red sash, moustache, and all! We froze.

He eyed us without expression. Then he growled: 'You must remember, careless talk costs lives. See you in the morning, lads.'

*

In fact, we didn't. He must have felt the shock he had given us was enough. It was. We never repeated that mistake."

Coldstream Guards

You're in the Book — No, As you WERE!

Guards' discipline is legendary. Deliberate breaches of that discipline bring down severe punishments on the guilty, and recruits quickly learn the standards they are expected to maintain and how to stay out of trouble.

There was one occasion, however, when a large number of men at the Caterham Depot breached the code — and got away with it.

Former Grenadier Vic Bradley (2627529), now living in Nottingham, recalled the day of the breakdown ...

"I was called up under the State of Emergency at the end of February, 1945, so I was still at Caterham doing my initial training when the war in Europe ended and VE Day was proclaimed. The Government decreed that it should be a public holiday, and all kinds of national celebrations and festivities were planned to mark the day and the end of hostilities here, before attention was concentrated on the war against Japan in the Far East.

As a gesture in keeping with the general mood of relief, the Depot CO gave staff and recruits the day off, the exceptions being the 'skeleton' force retained to do guard duties and maintain essential services. I spent most of the day in Purley. The lads whose homes were within reach — in London and the home counties — went off to join their families for the VE Day celebrations.

At roll call the following morning, most had not returned.

As a consequence, the Depot was half-empty when it should have been returning to a day of normal duties. Technically, the missing men were ABSENT and due to be dealt with under the standard code. A diplomatic solution was found. A second day off was declared.

It worked. The absentees returned within the day's grace and were hugely relieved to find they wouldn't face charges.

For one day, the Depot's strict discipline regime fell apart. But after four years of total war in Europe, the circumstances were exceptional. Leniency could be justified.

Things then quickly got back to normal!"

Caterham 'Crunchers'?

From the Croydon Chronicle, Mar 1892

———:o:———

The Reigate Priory men were due at Caterham on Wednesday to meet the Guards' Depot Football Club, but they did not turn up or assign any reason, it is said for their absence. Probably they were afraid to tackle the strong combination of Scots, Grenadiers and Coldstream - or was it the weather that prevented them? There is no sport which does anything like so much good for medical institution, in the way of providing medical practice, as football. And probably, had the match taken place, the medical staff of the Caterham Cottage Hospital might have had a little extra work in the way of bone-setting ...

———:o:———

A Dream's Prophecy

John Fletcher Boughey was one of the War Poets of WW2. And like Rupert Brooke in WW1, he foresaw his own death in action.

Born at Caterham on March 22, 1919, he was educated at Eton and graduated from Magdalene College, Cambridge. Then he came back to Caterham and was commissioned in the Coldstream Guards. He was just 21 when he was killed in action on August 31, 1940.

His verse, Last Night I Dreamed, is included in an Oxford University Press anthology of poetry written by young men from English public schools, who died in the war of 1939-45. Its title — For Your Tomorrow.

Last night I dreamed that to my bed there came
A heavenly vision, calling me by name
Who, standing in the moonlight, softly said
'Arise and follow me, for you are dead'.

Then in a sudden, timeless flash unfurled
Before my eyes, the compass of the world;
Before me lay eternity revealed
And all the secrets in earth's bosom sealed.

Thus, for a space I stood and watched, supreme
Wrapt in the matchless splendour of my dream,
And, suddenly I knew that time and death
Were but the fable on man's lying breath.

In the 1800s . . .

Surgeon-Major Sheldrake appears to have been a bit of a dandy

The IRA

A Terrible Night

Caterham Barracks came under attack twice in the century the Guards served there.

The first was an act of war, when the Luftwaffe bombed and machine-gunned the Depot at the outset of the Battle of Britain in 1940. It went on for many months, culminating in the random assaults by the first of Hitler's terror weapons, the V1 flying bombs. These days are described elsewhere in this book.

Thirty years later, a murderous attack — without warning and designed to kill Guardsmen and civilians alike — targeted the Caterham Arms pub, just outside the Depot perimeter. It was part of the IRA's bombing campaign of mainland Britain.

The 1st Battalion of the Welsh Guards was doing duty at the Barracks under their CO, Lieutenant-Colonel Peter Williams, on August 27, 1975. Now Brigadier Williams, he recalls that terrible night and its aftermath ...

"It was much like any hot summer evening. The clocks had just struck nine and the Depot was quiet. There was the usual 'walking out' and my wife and I were sitting at home in the old Commandant's House, reading.

Suddenly, there was a boom, quite unmistakable to anyone who had served in Northern Ireland but like all such noises, difficult to pinpoint initially. I ran at once to the Guardroom, to be met by the Guard and the kind of scene we had all witnessed in Belfast, now horrifyingly come to Caterham. Across from the barracks, smoke was billowing from the Caterham Arms, shards of glass covered the road and pavement, and dazed and bloodied men and women were staggering out gasping for air. Others with horrific injuries were carried out.

Looking back now, I give thanks for a curious quirk of fate which came to our help in this tragic situation. Nurses who lived out from the mental hospital next to the Depot sometimes returned of an evening to play bowls with the patients, the games finishing about 9pm. This was such an evening, and without hesitation they plunged into the crowd and began to administer first aid. Police, ambulance and fire services quickly reached the scene and restored order in their usual efficient way. The fires started by the explosion were extinguished, and the casualties evacuated to Redhill General Hospital.

Of the twenty Welsh Guardsmen taken to Redhill, eight were able to leave next day. Four of the others were on the Very Seriously Ill list and there were grave doubts about the survival of one, Guardsman Paul Thomas.

Forensic examination at the scene later indicated that the bomb had been placed under a table at which the most seriously injured had been sitting. Paul Thomas had been closest to it and lost his left leg below the knee, his right leg above the knee, and his right arm. Lance-Sergeant Stephen Ollerhead lost his right leg below the knee, Guardsman Garth Watkins his right leg below the knee, and Guardsman David Williams had fractures to his right leg, and severe blast injuries to both legs, his left arm and face. His lungs had also been badly scorched.

But they were all alive, thanks to the constant care and devotion of the surgeons, doctors and, nurses of Redhill General Hospital.

News of the bomb brought the national Press and TV to Caterham in force and the Adjutant, Captain REH David, and Orderly Room CSM Les Ellson braced themselves for a long night on duty in the Orderly Room, which quickly began to resemble an operational war headquarters rather than a peacetime office. My Second-in-Command, Major Charles Guthrie (now Chief of the Defence Staff) drove down from London that night to help cope with the media demand for information.

The Prince of Wales, the regiment's new Colonel, was in touch immediately and later came down to visit the injured in hospital. His presence gave all of us a lift, especially the injured and their families. Not surprisingly, there was an avalanche of letters of sympathy from friends and colleagues in the Armed Services. But we were all astonished and touched by the flood of public support, and letters from the ordinary people of this country, Canada and Australia, some with generous donations to the fund which had been established at once by the Mayor of Tandridge. The Army Benevolent Fund offered help on day one.

Guardsman Paul Thomas, whose terrible injuries had been life-threatening, was fighting back with the grit and determination of a rugby player from Neath. Indeed, within a few days he was holding his baby son, Nicholas, then one month old, in his good arm. His indomitable cheerfulness and determination to make the best of his life after suffering such injuries struck a chord with those who read newspaper reports of his gradual recovery. So much so that when the fund was wound up five months later, £70,000 had been raised — a very considerable sum in those days.

No amount of money can compensate for such damage and handicap, but it can help. The bonds that bind serving and former Guardsmen are strong and members of Association Branches of every Household Regiment responded magnificently. Less expected, but greatly appreciated, were donations from people who had organised sponsored events, pub quizzes, and events staged by telephone exchanges, rugby clubs and British Legion branches.

The British Limbless Ex-Servicemen's Association was also a great inspiration. Sir James Stuart-Menteth who, as a subaltern in the Scots Guards had lost both legs in action in 1944, visited to encourage and comfort, and show what could be done in life in the face of such circumstances.

Recovery took time, as well as the courage and strength of the injured. The process of claiming compensation through the Criminal Injuries Board was slow. But, gradually and painfully, it all worked.

Almost a year later, on July 30, 1976, the British Limbless Ex-Servicemen's Association organised a limbless climb of Snowdon. The climbers were accompanied by Prince Charles as Colonel, Welsh Guards. The Guardsmen who survived the Caterham bomb took part, notably Paul Thomas who manfully forced himself up the last hundred feet on his two new artificial limbs, after being lifted in by a helicopter of the Queen's Flight.

A generation on, a sense of perspective prevails. A few years later, the Battalion was to suffer terrible casualties in Bluff Cove, during the Falklands War. I went to see the wounded in hospital on their return. The thing which doesn't change is the spirit with which these disasters have been met. The determination, the self-discipline, the comradeship — even the humour of the injured and their fellow-Guardsmen — were the same as we had seen at Caterham.

So was the warmth of the response from the rest of the Army and the public at large. The lessons instilled in all Guardsmen at the Guards' Depot are the strands that make the cords that bind us all together in the priceless family that is the Household Division.

The spirit which helped Paul Thomas in his recovery is the same as that shown by Welsh Guardsman Simon Weston — another rugby player — in overcoming the horrific burns he sustained at Bluff Cove.

So, life moves on. In December 1997, Paul

Thomas proudly watched his son, Nicholas, whom he had cradled in one arm 22 years earlier, march up the steps of the Grand Entrance at Sandhurst — a newly commissioned officer in the Queen's Dragoon Guards.

I'm sure the shades of all the old Caterham instructors were smiling."

**Gerald Kersh
creator of
Sergeant Nelson**

Sergeant Nelson — and Hi-de-Hi!

The writer, Gerald Kersh, was a volunteer Guardsman during WW2. Squadded at Caterham, he spent his free hours writing about his experience there and, later, at the Pirbright training battalion. The result was the best-selling They Die With Their Boots Clean, followed by a film of the same name.

Kersh died in 1968, and the book is currently out of print. We are granted permission by the publishers, William Heineman, to include some extracts from the Caterham chapters. Gerald Kersh was a Coldstream (2663141), but his account of Depot life under the fictional 'Sergeant Nelson', his dawning understanding of the transformation his instructors were achieving, and his growing respect for them and his fellow squaddies, will be recognised by all Guardsmen, past and present. The title Kersh gave his book came from his Squad Sergeant:

'Now you mummy's darlings, get a rift on them boots. Definitely shine 'em my little curly-headed lambs for in our mob, war or no war, you die with clean boots on.'

Introducing Squad Sergeant Nelson:

There comes into the hut a man in shirt-sleeves and a soft SD cap. You can tell by his walk that he is no ordinary man. He swings his legs out from the hip, and his iron heels cut little arcs in the floorboards. He is long and lean, sun-dried, wind-cured, boucanned, smoked, and sandblasted. His face is brown as a kipper, and as expressionless. One of his eyes is fixed in a dreadful stare: it is of glass. The other blinks. There is nothing left of him but bone and sinew and vitals. Years of service have sweated away all that was superfluous or decorative. He has an air of demoniac energy: a wild swagger, a steady, genial ferocity. Out of his neatly-rolled sleeves hang arms as dark and gnarled as old salami sausages. He has fists like mallets of black stinkwood, an aluminium ring, and a silly little blue bird tatooed on his left wrist. Quite effortlessly, he shouts, in a voice that makes us jump:

'I am Sergeant Nelson! (Ain't I, Trained Soldier Brand?) I've got one eye, but both me arms! I died at Trafalgar but they dug me up again, and when I'm mad I'm a one-man wave o'destruction! I'm poison. I'm terrible. I kill seven rookies before breakfast! I can spit fifty yards through the eye of a needle! Do you see that dead tree over there? They'll tell you it was struck by lightning. Don't believe 'em. I killed it! I slapped it down! You're my new squad! I'm your Squad Instructor! Silence! Nobody say a word! You do as I say or you suffer. You suffer 'orrible tortures! Now when I say Hi-de-Hi Squad! you shout Ho-de-ho! — and shout it loud! Now: Hi-de-Hi Squad!'

We roar: 'Ho-de-Ho!'

'Right. Whenever I shout Hi-de-Hi, let me hear you reply pretty damn quick, or I'll chase you all round and round that square till the huts look like hen-houses.

'Hi-de-Hi!'

'Ho-de-Ho!'

'Good. Now we're introduced. I'm here to make Guardsmen out o' you.

. . . 'are you going to work with me?'

A chorus: 'Yes, Sergeant.'

He roars again. 'Okey-dokey, my little fluffy-'eaded chicks! Hi-de-Hi!'

'Ho-de-Ho!'

'Good. Now look. Recruits are babies. In one second, Cookhouse is going to blow. By rights I ought to march you about everywhere, definitely everywhere. But I'm going to let you go on your own, just to show I trust you. You won't get lost? Or make mugs of yourselves in any way?'

'No, Sergeant.'

Come to the cookhouse door, boys, cries the bugle.

'Knives, forks and spoons, and scram, then!'

We rush to the door.

'HALT!'

We stop, paralysed by that shattering voice.

'Hi-de-Hi, Squad!'

'Ho-de-Ho, Sergeant!'

We go to dinner.

Air-Raids:

... a bugle sounds, a siren moans, and TS Brand says: 'Jerry in the sky. Get in the trench.'

The Guards' Depot exploits air-raids, and makes prompt action a part of Guards' training. We run to cover, and it is then that Sergeant Nelson, who, for eight weeks, will never let us out of his sight, tells us about the Wogs, 'light of ear, bloody of hand,' the Arabs; and tells tall stories in short sentences, of discipline and training in the Guards. 'And if I make your blood run cold — don't worry, because I'll warm it for you when I get you on the square to-morrow. Definitely!'

On the Square:

'Now listen to me. I'm going to teach you some elementary drill movements. You don't have to be a Bachelor of Science to do 'em. Millions have done 'em before. Millions will do 'em again. You don't need a matriculation certificate to do it. Just be confident. Don't be nervous. Keep calm and do exactly as I tell you. And work! By God, work with me and I'm as mild as your mummys' milk. But work against me and I'll kill ya' ... Now for the time you're here, you'll shout out the time of your movements as you move.

'For example. Look at me. I'm standing properly at attention. See? Heels together, feet at an angle of forty-five degrees, fingers curled up, head perfectly still, eyes straight to the front, chin in, back like a ruler, and thumbs in line with the seams of the trousers. (That's what the seams of your trousers are for.) Now I'm standing to attention. I get the order Left Turn. **Le-heft . . .Tyeern!** The heel of my left foot becomes a pivot. I push with the toe of my right foot, and turn left. That's **One**. I count a pause — **Two, Three** — then raise my right knee smartly and bring my right foot down in the correct position at an angle of forty-five degrees with a smash that cracks the asphalt — **One ... two, three ... One!** Get it? You'll only shout out the time while you're here. By that time, the correct pause will become instinctive ...'

Nelson's Agony:

'Now, listen,' says Sergeant Nelson, dropping his voice to an ordinary, conversational tone. 'Some of you guys may be sensitive. I dunno. Well, don't worry if I shout at you a bit and call you names. It's essential. It's impossible to get along without it. I've got to get you fairly proficient in eight or nine weeks. Always be sure that I won't dish you out more than you can take, and I won't punish any man unless he asks for it. Take everything in good part, and you'll be all right.' He bellows: 'Now then! Come here. Lemme arrange you, like flowers in a garden ... oh, you pretty-pretty bunch o' soppy-stalked shy pansy-wansies. God definitely blimey, blimey with thunderbolts, blimey with lightning!' ...

To our left, thirty men, followed by a shrieking Sergeant brandishing a pace stick, execute a left wheel which, to us, represents the ultimate perfection of military footwork.

'God milk the coconuts and stone me over the hurdles,' groans Sergeant Nelson. 'Look at 'em. Three weeks squadded, and when it comes to a left wheel, some of 'em right wheel, some of 'em about turn, some of 'em turn handsprings, some of 'em pick their noses, and some stand still. Definitely horrible. Now you're going to show 'em what **you** can do!

With the possible exception of a man in love, no man in the world is so desperately eager to please as the new recruit in the Army. He has his back to the yardstick of regimental tradition. For the duration of his time, his value will depend upon nothing but his proficiency as a soldier. The muscles of a rookie doing his first Stand-at-Ease are as taut as those of a man clinging for his life to a breaking branch ...

Nelson on Saluting:

'Now, about saluting. They're pretty hot on saluting in the Brigade of Guards. I don't care what they do in the Boy Scouts or the Church Lads' Brigade or the WAACs or the ATS, or the Salvation Army. Here, saluting being a matter of discipline and proper courtesy and respect, they are hot on it, and rightly so. Thus, whenever you see an officer approaching you, you will salute ... head up, chin in, shoulders back, hand in line with the forearm, thumb pressed close against the edge of the hand, fingers all close together; the whole to come up like a steel spring, so that the right forefinger rests one inch above the right eyebrow' ...

Weeks into the Training:

'Sleep and refresh your pretty little selves,' says Sergeant Nelson. 'Because to-morrow I'm really going to start in on you. Definitely, I'm going to chase you to-morrow. I got a liver. And when I got a liver I'd tear my own grandmother's tripes out and trample them underfoot. I'd definitely do all that and much more.

Woho, am I going to chase you to-morrow! Any idle man here can make his last lousy little will and testament. Any chancer can go to the ablutions and cut his scraggy little throat from ear to ear into a wash-basin. Sleep! It's an order.

'Hi-de-Hi!'

We roar at the top of our voices: 'HO-DE-HO!'

We have suddenly become cheerful. We are getting the hang of things.

Progress:

Quickly, but smoothly, week after week, Sergeant Nelson drives his stuff into us; tireless, patient, with legs of steel and a throat of brass. Step by step he teaches us to march and drill ... we pass our second, fourth, and sixth week inspections ... Sergeant Nelson informs us that, in a long and varied life spent mostly among half-wits and the offscourings of the lunatic asylums of the earth, he has seen worse squads than us — which pleases us more than anything else ...

The Commandant's Inspection:

'Sergeant Nelson's Squad, eight weeks squadded, and ready for your inspection , **Sir!**' We stand frozen, stiff as overwound clockwork. We don't see the Commandant. We are staring straight to our front. We feel him as he passes . . . a presence, an eye. Seven or eight years pass. His voice is heard to say that our turnout is, on the whole, quite good. Sergeant Nelson's lone eye seems to heave a little sigh all on its own. The buttons on his SD jacket rise several inches, and then sink luxuriously. Then comes his voice. We thought we had heard him shout. We never did. He is shouting now . . . We stamp and wheel, right form and left form, salute to the left, the right, the front, and as improperly dressed.

We take up arms, and slope, and order, and trail, and present them. We feel that we are doing all right. The Commandant questions us. We have swotted up everything he is likely to ask us . . . He asks us all that and a lot more. The autumn goes, winter comes and goes, more summers come and fade. Years pass. Our beards are long and grizzled. Our eyes are rheumy with advanced old age. He will dismiss us, and then we can lie down and die.

We are dismissed. It took about an hour and a half ... 'How were we, Sergeant?'

And Sergeant Nelson says: 'You were not like Guardsmen. You were like a lousy crowd of wild, undisciplined Soudanese bloody fuzzy-wuzzies trained by illegitimate Wog corporals in a stinking pot-house in Tel Aviv in 1890. You were awful. I can never look myself in the face again. But all the same, you have passed out. . . '

We raise a wild cheer.

'I daresay you'll leave for the Training Battalion on Tuesday,' he says.

Topham Picturepoint

Gerald Kersh drew heavily on his Caterham and Pirbright experiences for the highly successful novels he wrote during and after WW2. His character, Sergeant Nelson, was a compound of the squad instructors who put Kersh through his early training. All former Guardsmen will recognise Nelson from their own days.

The Depot Pubs

The Fox

The Caterham Arms

Both pubs feature in the history of the Caterham Depot

In the 1800s . . .

From the Architect's (Major Sneddon) drawings.

Miscellany

THE RECRUIT
E.R
IRISH GUARDS
16/I
per week
&
ALL FOUND
6/9
per week
&
ALL FOUND
ENLISTMENT
18 AND 25 YEARS OF AGE
THE TERMS OF SERVICE are
3 years with the Colours and
9 years in the Reserve
21/-
per week
&
ALL FOUND
32/-
7/II
43/-
per week
& ALL FOUND
COMMISSIONED
OFFICER
TOWER WARDERS
COMMISSIONAIRES
PARK KEEPERS
POSTMEN
STEWARDS
GOD SAVE THE KING
II/5
per week
&
ALL FOUND
30/-
6d.

The Depot Chapel

The Chapel was at the heart of regimental and family life throughout the twelve decades the guards were at Caterham.

The Colours of the regiments, with their battle honours, hung proudly over the generations which attended services there. It saw countless baptisms, confirmations, marriages and funerals of members of the Guards' families. It was also the focus of prayer and quiet contemplation, away from the noise and bustle of Depot life, especially during wartime and other periods when Guardsmen were serving in the world's trouble spots.

It remains the Depot's only 'listed' building, a fact which gives it a measure of protection for the future.

William Butterfield, who designed the Caterham Depot chapel, was a Londoner, and a leading exponent of the Gothic Revival in architecture. He was born in 1814 and died in 1900.

Among the many famous buildings he designed were Keble College, Oxford; the chapel and quadrangle of Rugby School; St Alban's Holborn; All Saints, Margaret Street, London, and, in the antipodes, Perth and Melbourne Cathedrals.

In 1884, Butterfield was honoured with the award of the Gold Medal of the Royal Institute of British Architects. But he was a controversial figure and clearly there was more than one school of thought about his work. His obituaries demonstrated a few of the differing views of his career.

Manchester Guardian — 'The most original feature of his work was the stress that he laid on colour in architecture; he was fond of using coloured bricks and mosaics to vary the wall surfaces of his buildings, often it must be confessed, with singularly unhappy results. Keble College, Oxford, is Butterfield's most important work, and shows equally well the merits and the failings of his style.'

The Builder — 'The death of Mr Butterfield at the advanced age of eighty-five, marks the close of a noble architectural career, and takes us back in thought to the commencement of the Gothic revival, and the great modern church-building movement ... He was a man of far too individual and fervid temperament to miss putting into all he did something that especially stamped it as his own. And hence, while so many learned imitations of mediaeval churches ... are to the present generation but as dry bones, Butterfield's churches have still a life of their own ... Here is surely a lesson, and one of no little weight and importance, to the young architects of the present generation.'

*

Building the chapel began in 1886 and was completed early the following year. The War Office contributed £3,000 towards costs, but the figure had to be supplemented by public donations. Behind the altar is a mosaic showing Christ as King with the archangels Michael and Gabriel kneeling on either side.

The stained glass of the East Window shows the resurrection of Christ, with attendant groups of disciples.

The Colours of all five regiments hung in the chancel —

King's Colour, 4th Bn. Coldstream Guards, deposited 1919

Regimental Colour, 2nd Bn. Coldstream Guards, deposited 1936

King's Colour, 2nd Bn. Coldstream Guards, deposited 1936

Regimental Colour, 1st Bn. Scots Guards, deposited 1938

King's Colour, 1st Bn. Scots Guards, deposited 1938

Regimental Colour, 1st Bn. Grenadier Guards, deposited 1936

Regimental Colour, 3rd Bn. Grenadier Guards, deposited 1922

Regimental Colour, 2nd Bn. Welsh Guards, deposited 1959

King's Colour, 4th Bn. Grenadier Guards, deposited 1919

Regimental Colour, 2nd Bn. Scots Guards, deposited 1912

Queen's Colour, 1st Bn. Coldstream Guards, carried 1857-1872

Regimental Colour, 1st Bn. Coldstream Guards, carried 1857-1872

King's Colour, 3rd Bn. Grenadier Guards, deposited 1922

Regimental Colour, 2nd Bn. Irish Guards, deposited 1957

Chapel Interior with Colours

The Life of the Chapel

112 THE BRIGADE OF GUARDS' MAGAZINE.

S. Michael, Archangel, Guards' Chapel, Caterham.

Holy Communions.		Divine Service.		Baptisms, and Churchings.
Sundays.	Holy Days.	Sundays.	Week Days.	
8 a.m. on 1st, 3rd, and 5th Sundays, and after Parade Service on 2nd and 4th Sundays in the month. On Great Festivals at both 8 and 12.	8 a.m.	11.30—Parade Service. 3 p.m.—Children's Service on 1st Sunday in the month. 6.30—Evensong and Sermon.	Friday 7.30 p.m. Evensong and Address.	At any time after twenty-four hours' notice.

From Parish Notices — 1884

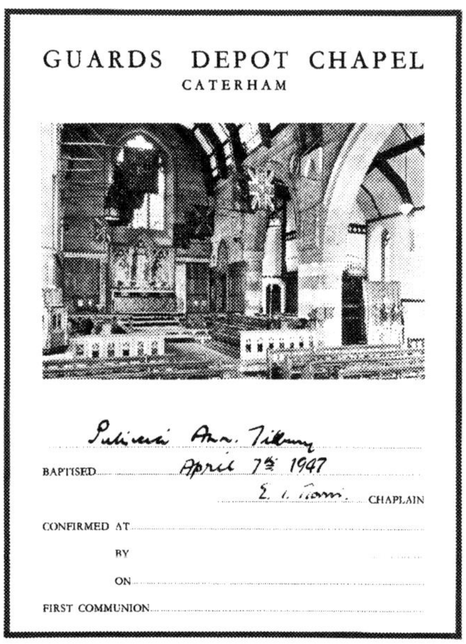

Guards Depot Chapel Choir.
SEPT. 1945.

L. Palmer, M. Kernick, J. Jenkins, K. Atterwill, W. Walker, K. Wright, H. Walter, K. Ackermann, J. Maworth, R. Dewey, R. Cooper
Mr. Bolton, Mr. Banks. Mr. Gray, C. Plumridge, J. Parker, R. Crabtree, G. Nott, P. Skinner, F. Booth, R. Ayliffe, D. Masters
Mr. Woods, Rev. J. ten Bruggenkate, Major E. Fryer, M.C., Rev. A. Naylor, O.B.E., D.S.O., Lt.-Col. G. Dury, M.C., Rev. E. Morris, Capt. W. Smith, M.B.E. Capt. T. Nation, Mr. Milne, Mr. Oates
(Verger) C.F. (Churchwarden) A.G.C., E. Command (Commandant) S.C.F. (Secretary) (Churchwarden) (Organist) (A.S.R.)
R. Harper, R. Nott, D. Nott, P. Vickerstaff, H. Tomlinson, J. Silvester
D. Johnson, D. Langley, T. Peters

Caterham's Characters

Some reminiscences by Barney Creton

Caterham in the early 1900s was quite small, but like the Depot it had its full share of characters — the individuals who stood out from the rest because they were 'different' in one way or another. Many of them had links with the barracks because they were tradespeople, small businesses which met the daily need for goods and services.

"I remember many of them. My mother gave birth to me in the Depot on the eve of the First World War. My father, Guardsman Frank Creton, was a 'pioneer' or blacksmith who had joined the Scots Guards from Glasgow at eighteen and was finishing his service at Caterham, after active service in the Boer War. He was proud of his long service medal and South Africa medal, with bars marking eight different actions in that campaign.

One of my earliest memories is being wheeled in a pram through 'tin town', the hasty, emergency accommodation put up for the huge influx of recruits when war came. My brothers and I knew all aspects of Depot life, and we came to know men who arrived at Caterham from all parts of the country. My brothers and I ranged through the town throughout childhood and youth.

Another early memory is of Mr Parr, the local barber. My father had got his first Army haircut from him, although Mr Parr had been cutting recruits' hair for many years before that. Dad said he was 'clever with the scissors' and took me and my four brothers to his shop for a regular trim. He wasn't in any way a 'demon' barber by reputation, but could guarantee to cut your hair in record time, drunk or sober.

His shop was in the High Street, handily opposite the Blacksmith's Arms pub and if he wasn't shaving or cutting hair when we arrived, we sat and waited for him to finish his pint across the road. Mr Parr had only one standard haircut for everyone — very short, but with a tuft left at the front. He used the tuft to turn our heads while he cut our hair. Dad never gave us any money for haircuts. He always settled up with Mr Parr in the pub. I believe it was threepence each.

When he wasn't barbering or enjoying a pint, Mr Parr would stand at the door of his shop with his little white apron and talk with all the passers-by. He had a ready wit and everyone knew him. He was a kind of hero to us kids because he knew a lot about a lot of things and enjoyed a laugh. For a long time we thought he was in the Guards, his connection with the Depot was so close. Later, we realised he couldn't have been. He wasn't tall enough.

Everyone in Caterham also knew Mr Taylor who kept a second-hand shop selling everything and anything capable of going through his doorway on Coulsdon Road near the Tally Ho! pub.

Dad got a gramophone from him and the records for us kids to play on it. Later, Mr Taylor was also able to provide a piano, side drum, guitar, crystal wireless set, and a variety of other items in keeping with the times.

It was a family business. Mrs Taylor also dealt in second-hand goods from several small shops in the town. He himself also had a 'transport' sideline, doing all kinds of jobs with his horse and cart, and his son, Oliver, ran the local coach service.

Then there was a Mr Gibson who kept a grocery and hardware shop, beside the Depot main gate on the Coulsdon Road. His son, Alan, did the deliveries in the Depot and through the town, riding his father's delivery bike with the big basket on the handlebars. Apart from food, one of his main deliveries was paraffin for lamps — it was before the days of gas and electricity. I remember we all thought it was absolutely marvellous, the day when Dad told us we were going to have gas for lighting and cooking. We got taps on the wall which we had to open when it was time to light the gas mantles. It meant a big drop in Mr Gibson's paraffin sales. But we still needed some for Mum's incubator for duck and chicken eggs!

Alan Gibson was a keen cricketer and was a frequent visitor to the sergeants' mess in the Depot where his interest in the game made him welcome. He didn't look out of place there because he stood over six feet tall.

Another local worthy was Mrs Manning. She was a regular at the Tally Ho! pub, down the road from the barracks. To everyone, including many Guardsmen, she was a familiar figure in a style of dress as old-fashioned as she herself was. She always arrived at the pub in her pony trap, emerging from Cromwell Road, just opposite, at about the same time each day. Joe Waylock, Tally Ho! landlord, and Stephen Penney who succeeded him, always made sure she never came to harm. Many a Guardsman had to help Mrs Manning into her trap at the end of the evening, but nobody ever laughed at her. To everyone then, she was just a grand old lady, quiet and respected, even although she might take a little more refreshment than was good for her. Her pony was totally reliable. Pointed in the right direction, he would end up in her garden where he would wait patiently at the stable door for a daughter to come and take his mistress safely indoors.

Another well-known cricketing family in the early 1900s were the Wilsons. Best known was Barney Wilson because he played for Surrey and he, too, was a familiar figure in the Guards' Club. His father kept a little newspaper shop on what recruits knew as Burns Common.

Another Caterham shopkeeper of these days was old Mrs Bolton who kept a sweets, cigarettes and small provisions shop near the Tally Ho! She wore very small glasses over which she would peer at youngsters as she weighed out a 'ha-porth' of sweets, or a farthing packet of sherbet with a liquorice tube stuck in the top for us to suck it up. Or it could be a few ounces of cheese or butter for Mum. Her son, Victor, in a spotless white apron, stood at the door to make sure we didn't pinch anything.

I developed my passion for cod and chips back in the old days at the fish and chip shop run by Mrs Edwards near the Depot gate. These marvellous meals she served up are the reason, even at 87, I cannot resist cod and chips. It's still my favourite.

Mrs Perks, whose husband retired from the Depot, made her home into a general store not far from the main gate by taking out the front window of her terrace house and replacing it with a counter. They used to say you could buy almost anything from Mrs Perks.

Then there was old Mrs Knight who kept what was called a 'dairy' with a gold-coloured cow and calf in a glass case in the window. She sold milk and bread, delivered fresh from Wallington daily in a round-topped cart drawn by two high-spirited horses, with a bowler-hatted driver as high-spirited as his horses. It was a pleasure to run up to Mrs Knight's for a farthing loaf with a beautiful crust on it.

These were some of the trades-people who served the Depot and surrounding area. Those in married quarters knew them best, but young soldiers also got to know the locals fairly quickly. There was a community atmosphere which was helped by the concerts the Guards' bands would give at garden fetes and on holidays. These were very popular with local people. Military music has always been very entertaining to me!

It may strike the reader that trades-people, in business in a small way in that earlier, quieter time were known and referred to as Mr. This or Mrs. That. They were given the respect and recognition which what they did in the community, deserved from the community. Youngsters who didn't follow their parents' example in this attitude would have heard all about it pretty quickly. How different to-day!

The Cretons are a Guards family. My mother's sisters — Sarah, Lizzie, Beatrice, Norah and Cecilia — all married into the Brigade. My uncles were Guardsman William Guthrie, Scots Guards, whose son was a piper in the regiment; Sgt J. Nicholls, Grenadiers; Guardsman A. Fremantle, Scots Guards; Guardsman Walter Bland, Welsh Guards; and Guardsman Jimmy Exley, Coldstream Guards. My oldest brother also joined the Scots Guards. Now 88, he lives in Bath.

Our family tree grew a lot of its branches at the Depot. Now, I and many others miss the old Guards Depot. Looking back, I think it made Caterham . . . "

Grenadier Guards

Scots Guards

Coldstream Guards

Irish Guards

Welsh Guards

A Soldier's Garden

The gardens of serving soldiers tended to alternate between flowers in peacetime and vegetables in times of war. Like the allotments and gardens of civilians, soldiers' efforts shifted from the decorative to the practical when hostilities forced the introduction of ration books.

An example of peacetime gardening at Caterham Depot comes from an unlikely source, the magazine, American Homes & Gardens. The issue of November, 1911 featured the efforts of a Corporal Holt of the Coldstream Guards who had just won nine first prizes for his vegetables. Corporal Holt, claimed the magazine, was distinguished in two ways. He was the tallest man in the British Army as well as its champion vegetable grower. His garden was judged to be the best-kept in the barracks, as well as its best-cropped.

The magazine made comparisons between soldiers' gardens on both sides of the Atlantic, but concluded that the Americans' were far behind Caterham's.

———oOo———

Guards' Humour

The Miracle

... And one who was called Sergeant came unto them as they stood together, crying in a loud voice: 'Squad, s q u a d - SHUN!' And, even as he spake, it was so ...

———oOo———

In the 1880s . . .

An enigmatic sketch titled 'Shipping the Cargo' in which the Adjutant 'takes the Quartermaster for a drive'. Was there a Tumbledown Dick Pub in Caterham? The name originated with Richard Cromwell, son of the Protector, an 'unsteady' character.

Educating the 'Daughters of the Regiments'

Beechhanger Court, Caterham, was the Brigade of Guards' Boarding School for Girls between 1934 and 1946, educating 'daughters of the regiments' without charge to their parents.

It moved to Harestone Lane, Caterham, from a school in Francis Street, Westminster, which had been built for the purpose in 1865. It closed when new legislation on education, introduced after WW2, rendered it unnecessary. About 40 girls were in residence at any one time, educated, fed and clothed free. They were distinguished by the cloaks in the Brigade colours of red and blue which they wore on formal occasions at the School and Depot. The School's committee of management consisted of former officers, plus wives and mothers of serving members of the Brigade.

The School's annual prize-giving was held in July and was always attended by the Depot Commandant, at least one Regimental Lieut.-Colonel, the Depot Chaplain, and many parents and friends of the girls and their families. As well as the standard curriculum subjects, the girls were taught piano and singing, and often gave mini-concerts featuring music, gymnastics and short plays, at both School and Depot. These were always very popular events in the Brigade 'family' and often featured one of the regimental bands.

In one of the School's final years, fifteen girls received prizes for music, art, needlework and general subjects; eight girls won pianoforte certificates from the Royal School of Music, London; three received certificates for solo singing, and one for 'pianoforte playing', awarded at the Coulsdon-Merstham-Purley Music Festival. At the same festival, the School's junior choir won an Honours Certificate.

A Nostalgic Return

Many former Guardsmen have been back to Caterham since their service days, such has been the nostalgic pull of the old Depot for thousands of men who volunteered or were conscripted there into the hard, respected fraternity of the five regiments which, with The Life Guards and Blues and Royals, comprise the Queen's Household troops — the Sovereign's bodyguard.

Although Caterham ceased to be the Guards' Depot in 1960, Guardsmen have served there well into the '90s, keeping a holding presence in the shrunken barracks and maintaining its facilities until the Ministry of Defence finally decided that, after 120 years, it could be relinquished and given over to civil purposes.

Men of the Bristol Branch of the Coldstream Guards Association went back to Caterham with their wives in 1994. Branch secretary Reginald G. Counsell recalled:

"Most in our party had not been back to the Depot in 50 years and were unprepared for the change which confronted them. Nevertheless, we were given a cordial welcome by CSM Eddie O'Keefe and two Sergeant guides, acting on behalf of the RSM of the 2nd Battalion, Grenadiers, the resident unit. It was with a degree of sadness that we began this final tour.

Opposite the guardroom, like a symbol of another era, stood the main gate sentry box — empty, forlorn, redundant. Two Guardsmen in battle fatigues, with weapons reminiscent of outer space films, stood casually alert at the lowered barrier. In the guardroom, the duty sergeant sat at a desk, while the corporal of the guard watched four monitor screens fed by closed circuit cameras, which covered all outdoor activity around the Depot area.

Remembering the days of RSM 'Tibby' Brittain in a Depot pulsating with activity, we were struck by the silence and peacefulness of the place. At this time on a similar Sunday morning fifty years ago, five hundred conscripted souls would be raising the Chapel roof, and the Depot would be a hive of activity. Where the wartime slit trenches had provided cover from hit-and-run Luftwaffe raids, were well-tended flower beds.

After tea in the sergeants' mess, our guides gave us a tour of the barrack blocks. At Monck Block, we gathered for photographs and exchanged 'do-you-remember-when' stories about the old days. But the contrast between then and now was all around, in signs like 'Launderette' and 'NAAFI Automatic'.

No matter. It was a lovely day and the sun shone as we were escorted to the cricket pavilion. In our service days, we were never allowed near this hallowed spot. But to-day, we represented part of the Depot's past and our hosts graciously gave over the balcony to us. In the shade, we sat with liquid refreshments and watched the Grenadiers play the summer game.

Between overs, our gaze wandered to the backdrop of grim, grey buildings and the thoughts of elderly men turned again to their youth and how, in this place, they were abruptly (but never cruelly) turned from raw, undisciplined recruits into a special kind of men. Lifelong friendships and comradeship were moulded here, and bonds forged between men and regiment which no words can convey.

At lunch, our hosts laid on an excellent barbecue. While the ladies chatted and revelled in the warmth of the day and our reception, the men continued to roll back the years, recalling the host of Depot personalities who gave the place its character and made it what it was.

All too quickly, the afternoon and the match ended, stumps were drawn, and the teams left the field to applause and shouts of 'well played'. And that's the kind of day it had been — nostalgic and memorable. 'Well played' seemed to us to be a fitting, final tribute to the old Depot.

At the main gate we thanked our hosts for their hospitality and turned for a last look, half expecting to see the huge, unsmiling figure of 'Tibby' Brittain watching us.

We didn't see him. But I'd bet that he was there somewhere ..."

In the 1800s . . .

A Lieutenant Monck, Coldstream Guards, who seems to have spent his leisure hours studying the Army List as a guide to likely promotion prospects . . .

Memories of Marching Men, Cricket — and Young Love

by Mrs. Peggy Jones, Craig Cefn Park, Swansea.

My husband, 2737966 Sgt W.E. Jones, Welsh Guards, served as a PT instructor at Caterham for four years during World War II. He had great pride in the Depot and the men he helped to convert from rookies into trained soldiers.

In 1944, while I was Walt's fiancee, he arranged for me to stay with a Welsh family in Coulsdon for a week so that we could be together and I could get a glimpse of his life at the barracks. On the first day of my visit, he took me to watch squads drilling with so much precision on the square. I was very impressed. He said: 'By the way, Peg, would you mind if I played cricket to-night? It's an important match. You can watch, and I'll find company for you to sit with.'

I didn't mind and was proud to watch him bowling, and skittling the opposition out. However, I was less pleased later when he told me the match had been the first round of the inter-regimental competition.

And my initial pride went even thinner when the Welsh Guards went on winning through and finally took the trophy. I had spent four evenings of my week watching cricket at the Depot. It had to be love!

Sadly, Walt died three years ago aged 77. But these happy days at the start of our life together come flooding back when I look at my china cabinet, and the beautiful silver cruet inscribed with our initials and 'From the Guards Depot Cricket Team on the occasion of your marriage, September 1944.'

I salute the old Depot and its bustle and shadowy figures of marching men. For me, it means my beloved Walt whose world it was for four years.

A Widow's Plea

One thought which arose repeatedly from the contributions to this book was the wish that, whatever civil purposes the old Depot area should serve in the future, there must be recognition of the part it played in defending this country's interests and security for over a century — in particular, respect and honour for the thousands who died, having joined the Guards regiments at Caterham.

An elderly widow, Irene Small, who lives in sheltered accommodation at Caterham Hill, wrote to the chairman of the Barracks Development Group at Tandridge District Council to press the point. She speaks for the many:

"My husband was conscripted during the Second World War and joined the Scots Guards at Caterham. He fought in North Africa, then moved on with his battalion to Italy in the campaign to liberate Europe from the Nazis.

My brother, Norman Frederick Kitchener Palliser, ran away from our home in Ireland at the age of 17 and joined the Irish Guards at Caterham. My mother tried to buy him out, but he had already taken the King's Shilling. Eventually, both he and my husband were taken prisoner in Italy and went to prison camps in Germany. My brother had also served in Egypt and compiled an album of recollections while he was there. When he died in a drowning tragedy two years after the war ended, I sent the album to the Imperial War Museum where it is to-day.

I believe that the men who did their bit for Britain in a very dangerous century, and particularly those who didn't come back to enjoy the peace they had won, should never be forgotten.

Britain, and Caterham, owes them that."

Two Mysteries — Two Solutions

A Pyrrhic Victory . . . and a Cross that 'Vanished'

To former Grenadiers saying their final farewells at the old Depot, it was a mystery which couldn't be solved by the Ministry of Defence staff charged with security and the hand-over to the site developers. A revered memorial to the regiment's casualties in one of the bloodiest actions of WW1 had stood near the Grenadier Orderly Room since 1918. Now it had disappeared ... and nobody at Caterham could say what had happened to it.

The memorial marked a costly victory the regiment gained in the 1916 hell of the Somme. The Grenadiers were ordered to advance and attack a position held in strength by the Germans. Twelve times they attacked, only to be driven back by withering fire from the enemy positions. Despite heavy casualties, the Grenadiers drove forward again - and their thirteenth determined assault finally overran the German trenches and gained the objective.

A memorial was created to honour the 28 officers and 700 other ranks who died in the action. A large wooden cross, carved in heavy oak taken from the ruined church in the village of Saillisel, it was painted black and stood for decades in the Depot at the centre of a rosebed. Flowers were laid at the cross on each anniversary of the Somme battle.

In 1953, experts were asked to examine the Cross. They established that its wood was at least a thousand years old.

Then, as the old Depot was about to close, questions were raised about the fate of the Somme Cross. Men who had served with the regiment wanted to know that it still honoured the Grenadiers who had won such a bitter victory.

Researches for this book revealed that the Cross had never, in fact, 'vanished'. It had been taken from the rosebed some time before the closure date and repositioned beside the altar of All Saints' Church, Alexander Barracks, Pirbright, where it stands to-day.

The Somme Cross, where the Grenadiers placed it in the old Depot after WW1
Photograph by HW Hilton of Mundlesly, Norfolk.

Mystery of the Depot Guns

Two German Howitzers — WW1 trophies first displayed near the Caterham Depot's main gate in 1919 — disappeared in the 1930s. For years there has been speculation about their fate. Were they located elsewhere? Had they been sold for their scrap value? Or what? Nobody seemed to know.

Certainly, there was photographic evidence that the guns had been in the Depot, but inquiries yielded nothing more. Then a strange story began to surface. It was rumoured that the guns had been buried at the Depot, although nobody seemed to know why ... or where.

Then former Grenadier Bill White of Bury St Edmunds, who served at the Depot in the thirties, came up with an eye-witness account of what happened to the guns.

"After the end of the Great War in 1918, a good many cities, towns, military establishments, and the like, acquired one or more trophies of that dreadful period — usually a tank, or one or two guns.

The Guards' Depot at Caterham had two large guns, one displayed near the main entrance and the other in the vicinity of the Clock Tower.

I think it was in the early part of 1934, when I was serving at the Depot as a Trained Soldier, that I heard that the guns were being dumped. It was a strange situation and I was intrigued. I got to the Clock Tower in time to see the gun kept there being drawn away towards the back gate by Blossom, the Depot horse (with her cart, she was the only transport there at the time), assisted by a team of recruits.

Beyond the Back Gate, there was a large area of rough open ground. The gun was drawn, with difficulty, more or less in a straight line for about seventy to a hundred yards, to a spot where a deep pit had been dug. At that point, Blossom was unlinked from the gun, and the recruits took over. The first gun was already in the pit, and the recruits tipped the second in to join it. Both ended up more or less upside down and there, I suppose, they lie to this day."

Why the guns were buried is still a mystery. Photographs of the guns were sent to the experts at the Woolwich Arsenal in London. They identified them as German field pieces from WW1 — the larger one a heavy 15 centimetre howitzer, the smaller a 10.5 centimetre light howitzer. Both were almost certainly made by the German armaments company, Krupps. The most likely explanation is that the guns were deteriorating under exposure to the elements and were becoming an embarrassment rather than trophies, to be valued.

There is a sad footnote to this story. William E. White, Grenadier 2613167, served twice with the regiment — from April, 1932, to March 1936, and from December, 1939, to September 1940. He died, aged 85, in September, 1997, not long after he contributed his solution to the mystery of the German howitzers.

The Farewell

The Brigade Shield

Presented to Caterham when the Depot moved to Pirbright, the Brigade Shield hangs today in the town's Soper Hall.

The Tandridge Link

Tandridge's chain of office bears a link with the Brigade Star, 1877-1960

The Farewell

So an era ended, and Caterham ceased to be the Guards' Depot in 1960. The move to Pirbright had been dictated by the need for economies and access to large-scale firing ranges for modern automatic weapons. However, the presence of the Sovereign's personal troops continued for another thirty-five years. Various battalions of the five regiments continued to be based in the old barracks, although they occupied only a small and shrinking area where, in periods of national emergency, thousands had been under training simultaneously.

The relationship between the Guards and the local community was demonstrated in a formal exchange in 1960. Lieutenant-Colonel VF Erskine Crum MC, CIE, presented an additional link to be added to the Council Chairman's chain of office. It was engraved with the badge of the Guards' Depot.

In return, the Council Chairman presented a silver salver to the Brigade, inscribed 'Caterham 1877-1960. Presented by the Caterham and Warlingham Urban District Council on the occasion of the transfer of the Guards' Depot to Pirbright, March 1960'. With it, was a framed copy of the Council's unanimous resolution recording their 'regret at the departure, their pride in the long and happy association which the district has enjoyed with the Depot ...' and ... 'their appreciation of the valuable contribution made by the Depot over the years to the community life of the district.' On behalf of local residents, the Council wished the new Depot continued success.

In the Soper Memorial Hall, Caterham, alongside the badge of what had been RAF Kenley, is the badge of the Brigade of Guards, inscribed: 'This plaque was presented to the Caterham and Warlingham Urban District Council in July 1960, by all ranks of the Guards' Depot, as a mark of respect to the citizens of Caterham who have always made generations of Guardsmen welcome during the many years the Depot was stationed in Caterham'.

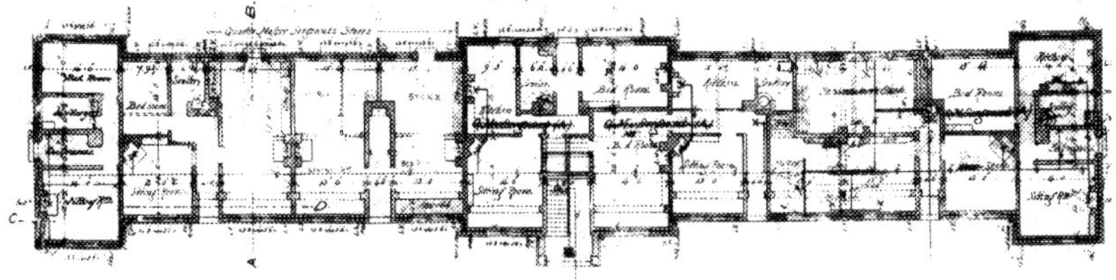

PLAN OF GROUND FLOOR.

. . . and The Long Goodbye

Although Caterham ceased to be the Guards' Depot in 1960, the five Foot Guards regiments of the Household Division continued to have a presence there till 1995. One by one, the Grenadiers, Coldstream, Scots, Irish and Welsh maintained battalions at the barracks in between spells of duty elsewhere.

This was the thirty-five-year farewell:

1st Welsh Guards	—	Jan '60 - Dec '60
1st Irish Guards	—	Dec '60 - Dec '61
2nd Grenadier Guards	—	Dec '61 - Oct '64
2nd Scots Guards	—	Oct '64 - Jan '66
1st Grenadier Guards	—	Jan '66 - Apl '70
1st Irish Guards	—	Apl '70 - Nov '70
2nd Grenadier Guards	—	Nov '70 -Dec '72
1st Irish Guards	—	Dec '72 - Jan '75
1st Welsh Guards	—	Jan '75 - Jan '77
2nd Coldstream Guards	—	Mar '78 - Dec '80
1st Coldstream Guards	—	Dec '80 - Feb '86
2nd Grenadier Guards	—	Mar '88 - Dec '94
Nijmegen Coy Grenadier Guards	—	Dec '94 - Mar '95

Apart from two short breaks when the Guards' battalions were fully committed elsewhere — January 1977 to March 1978, and February 1986 to March 1988 — there was only one other period after the commissioning in 1877 when there were no Guards units at the Caterham Depot that was early in the 20th century when all five regiments moved out for a year because of a serious outbreak of diphtheria in the area.

The Grenadiers' Nijmegen colour comes down — and the 118-year link is finally broken.

Finale

It fell to the Grenadiers to say the final farewell.

On Thursday, March 23, 1995, the flag of the Regiment's Nijmegen Company was lowered for the last time and the Guardsmen marched out, 118 years after the Depot first came alive. Groups of people watched them go as they marched down Church Hill and out of Caterham.

Then it's Tommy this, an' Tommy that
an' 'Tommy, 'ow's yer soul?'
But it's 'Thin red line of 'eroes'
when the drums begin to roll ...

Kipling's reproach - his criticism of the wavering civilian attitude to the British soldier - had no lasting justification in this garrison town. But two decades after he wrote them, he must have thought of Caterham and recalled the verses as he faced the tragedy which clouded the rest of his life.

The drums had begun to roll again in 1914. His only son, Lieutenant John Kipling, Irish Guards, was killed at Loos, one of the bloodiest battles of WW1. He was 18.

Silent Reminders

Many Guardsmen married local girls during their service, returning from foreign postings or on demobilisation to settle on the Hill or in the Valley. Their descendants still live in the town and honour their family links with the old Depot.

Although no longer a town with a Guards' presence, Caterham will always have something of the Brigade's traditions. As promised, the developed site of the barracks will retain reminders of its past. So will the graveyards of the town. Thousands of men who began their service in Caterham lie in military cemeteries overseas. But a hundred and fifty-six of the men who came back to live in the town are buried in St. Mary's Churchyard and the Caterham Burial Ground, interred there since the middle of the 19th century.

The Standard-Setters

The Men Who Ran the Depot

The life of the Caterham Depot and the pursuit of excellence in its training regime always depended on teamwork by its staff — Officers, Warrant Officers, NCOs and Trained Soldiers. The standards which the regime achieved consistently for over a hundred years bears testimony to the breadth and depth of the Brigade's resources in its training personnel.

Like the continuous intake of recruits, the Depot staff was itself a shifting population. The staff at all levels were posted to Caterham from their battalions as they alternated between home and overseas postings. They served their period in whatever aspect of training they were selected for, and were posted out again as their replacements arrived. A few served long periods; most were in and out again in two or three years. To the recruits, it was a seamless process from which they moved on to the next stage of their training in a battalion.

It is a fact that till the end of his days, every Guardsman could tell you the names of his Caterham squad instructor, trained soldier and officers as readily as he could his own regimental number. These were men whose impact was never forgotten. Even if we could identify them all, they would be too numerous to name here. However, available records and space allow us to name two categories of Caterham's senior staff — its Commandants and Regimental Sergeant Majors.

The Commandants

Major & Lieut.-Colonel C. D. Thomas	Coldstream Guards	Sept 1875 - Mar1882
Capt & Lieut.-Colonel W. H. Mackinnon	Grenadier Guards	Mar 1882 - May 1884
Capt & Lieut.-Colonel L. D. Mackinnon	Coldstream Guards	June 1884 - Feb 1885
Major & Lieut.-Colonel A. H. M Morton	Coldstream Guards	Feb 1885 - Nov 1885
Major E. Crabbe	Grenadier Guards	Nov 1885 - July 1890
Major E. E. Colville	Grenadier Guards	Sept 1890 - Sept 1891
Major H. Crauford	Grenadier Guards	Oct 1891 - Dec 1894
Major The Hon. A. H. Henniker	Coldstream Guards	Jan 1895 - July 1896
Major F. Lloyd	Grenadier Guards	Aug 1896 - July 1897
Major C. A. A. Frederick	Coldstream Guards	Aug 1897 - May1901
Major The Hon. J. H. Bailey D.S.O.	Grenadier Guards	June 1901 - Feb 1903
Capt & Bt. Major J. McSteele	Coldstream Guards	March 1903 - Feb 1905
Major J. Ponsonby D.S.O	Coldstream Guards	March 1905 - Feb 1907
Major F. de K. Steuart	Scots Guards	March 1907 - May 1908
Major G. P. T. Fielding D.S.O.	Coldstream Guards	June 1908 - May 1911
Major G. D. Jeffreys	Grenadier Guards	June 1911 - Aug 1914
Major H. W. Ingleby	Scots Guards	Aug 1914 - Mar 1919
Major L. M. Gregson C.B.E.	Grenadier Guards	Mar 1919 - Jan 1921
Major (Bt. Lieut.-Colonel) R. E. K. Leatham D.S.O.	Welsh Guards	Jan 1921 - Jan 1923
Major Sir Victor Mackenzie Bt. D.S.O. M.V.O.	Scots Guards	Jan 1923 - May 1924
Major A. F. Smith D.S.O. M.C.	Coldstream Guards	June 1924 - April 1927
Major F. Longueville D.S.O. M.C.	Coldstream Guards	May 1927 - Sept 1928

*Major the Viscount Marsham	Coldstream Guards	Sept 1928 - Oct 1931
Maior M. F. Trew	Coldstream Guards	Oct 1931 - Oct 1934
Major W. D. Faulkner M.C.	Irish Guards	Oct 1934 - Sept 1937
Major A. L. Cazenove M.C.	Coldstream Guards	Sept 1937 - Sept 1939
*Lieut.-Colonel the Earl of Romney	Coldstream Guards	Sept 1939 - Dec 1943
Lieut.-Colonel G. A. I. Dury M.C.	Grenadier Guards	Dec 1943 - May 1946
Lieut.-Colonel J. C. Windsor-Lewis D.S.O. M.C.	Welsh Guards	June 1946 - Nov 1948
Major Sir C. H. F. Maclean Bt.	Scots Guards	Nov 1948 - Jan 1949
Lieut.-Colonel J. Chandos-Pole	Coldstream Guards	Jan 1949 - July 1950
Lieut.-Colonel P. F. I. Reid	Irish Guards	Aug 1950 - Sept 1953
Lieut.-Colonel I. W. Gore-Langton M.B.E	Coldstream Guards	Oct 1953 - Sep 1955
Lieut.-Colonel F. J. C. Bowes-Lyon M.C..	Grenadier Guards	Sept 1955 - July 1957
Lieut.-Colonel V. F. Erskine Crum C.I.E. M.C.	Scots Guards	July 1957 - Jan 1960
Lieut.-Colonel G. G. Fowke	Welsh Guards	Mar 1960 - Mar 1961

*Viscount Marsham, afterwards the Earl of Romney, was the only officer to become Commandant twice.

The Regimental Sergeant Majors

RSM H. Darrell	Coldstream	Jul 1881 - Aug 1881
RSM S.White	Coldstream	Aug 1881 - Jun 1883
RSM E. Dickerson	Coldstream	Jul 1884 - Apl 1885
RSM J. Jones	Grenadiers	Aug 1885 - Nov 1887
RSM H. Martin	Coldstream	Jan 1888 - Mar 1891
RSM A. Telfer	Scots	Mar 1891 - May 1891
RSM J. Sparkes	Coldstream	May 1891 - Jan 1893
RSM W. Stewart	Scots	Feb 1893 - May 1896
RSM W.J. Cock	Grenadiers	Mar 1897 - Apl 1897
RSM G. Gooding	Coldstream	May 1897 - Nov 1897
RSM J.C. Mant	Scots	Dec 1897 - Dec 1899
RSM J. Skidmore	Grenadiers	Sep 1900 - Feb 1901
RSM J. Boyd	Coldstream	Feb 1901 - Sep 1902
RSM J. Mitchell	Scots	Dec 1902 - Feb 1903
RSM C.A. Baylis	Irish	Mar 1903 - Jan 1905
RSM J. Teece	Grenadiers	Jan 1905 - Dec 1909
RSM F.J. Caunell	Coldstream	Dec 1909 - Apl 1911
RSM J. Tate	Scots	Apl 1911 - May 1913
RSM J. Kirk	Irish	May 1913 - Dec 1913
RSM H. Wood	Grenadiers	Dec 1913 - Aug 1915
RSM E. Ellis	Coldstream	Aug 1915 - Apl 1916
RSM J. Barwick MC	Scots	Apl 1916 - Jul 1919
RSM A.E. Pettit MC	Scots	Jul 1919 - Dec 1920
RSM C. Harradine DCM	Irish	Jan 1921 - Apl 1922
RSM W.E. Hawkins	Grenadiers	May 1922 - Oct 1925
RSM F. Gill	Coldstream	Oct 1924 - Apl 1930

There was an outbreak of diphtheria at the Depot in 1927, resulting in a temporary evacuation of Caterham Barracks.

RSM W.A. Blakeley	Scots	Jul 1930 - Jul 1931
RSM J.D. Hughes	Welsh	Jul 1931 - Feb 1934
RSM G.F.G. Turner	Grenadiers	Feb 1934 - Nov 1934
RSM R. Brittain	Coldstream	Dec 1934 - Oct 1935
RSM A.J. Brand	Grenadiers	Oct 1935 - Aug 1936
RSM M. Jones	Coldstream	Sep 1936 - Feb 1937
RSM W.J. Dorman	Scots	Feb 1937 - May 1939
RSM J. Hastings	Irish	May 1939 - May 1939
RSM H. McKinney	Irish	May 1939 - Jul 1940
RSM J.A. Stack MC	Irish	Jul 1940 - Sep 1940
RSM P. Dunne	Welsh	Oct 1940 - Jul 1944
RSM D. Hobbs MBE	Grenadiers	Jul 1944 - Aug 1945
RSM S.M.Hamilton MBE	Scots	Aug 1945 - Aug 1946
RSM W. Rooney MM	Irish	Aug 1946 - Dec 1949
RSM B.E. Hillier DCM	Welsh	Dec 1949 - Sep 1951
RSM W.L.A. Nash	Grenadiers	Sep 1951 - Mar 1953
RSM R.J.S. Tyacke	Coldstream	Mar 1953 - Oct 1953
RSM D. Whyte	Scots	Oct 1953 - Apl 1956
RSM L.C. Drouet	Grenadiers	Apl 1956 - May 1957
RSM F. Clutton MM	Grenadiers	May 1957 - Mar 1958
RSM W. Rodger	Scots	Mar 1958 - Oct 1960

The Guards' Depot moved to Pirbright in 1960.

Only the Shades of Old Guardsmen Parade Caterham Now . . .

The Barrack Blocks

The Square and Married Quarters

Drill Sheds

'Coronation Street'

. . and Guard Room

A Cushy Billet?

From the Croydon Chronicle, Mar 1892

———oOo———

What a difference there is in the treatment of the soldier of the present day contrasted with, say 30 or 40 years ago. The only amusement for the recruit then was to be found in the canteen, where the vilest beer and gin was quaffed, and the young recruit was compelled to listen to songs that would have fetched, not only M'Dougall's head gear off, but his beard and moustache as well. Now the soldier is provided with regimental libraries, rending rooms and free entertainments. At the Caterham Depot, everything that is possible is done to brighten his three or four months' residence, when undergoing training.

Ah, well — maybe our memories play us tricks . . .

Editor.

The Future

In the 2000s . . .

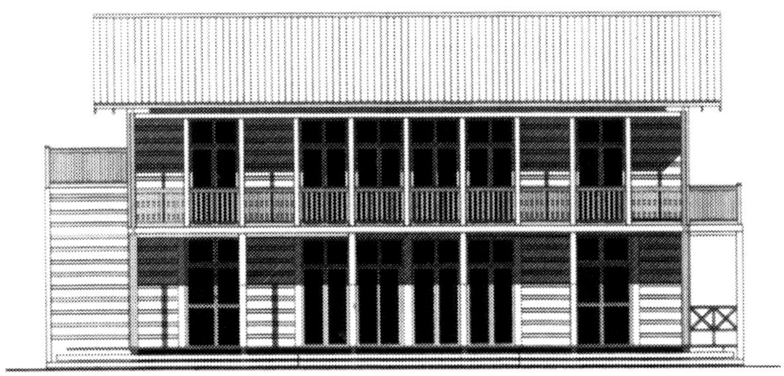

Front elevation of the proposed new sports pavilion

The Future

With the old Caterham Guards' Depot closed and its silent barrack blocks, drill square, chapel, sports fields, mess halls and gymnasia handed over for development three years after the last Guardsman marched out, the frequently-expressed hope is that its part in many turbulent decades and times of great national danger will not be forgotten.

The wish that a means will be found to remind future generations what happened here has been expressed by the people of Caterham as well as the men of the five regiments of the Guards Brigade, evidence of the good working relationship which grew between the community and the Depot.

More than a thousand local people and many former Guardsmen returned to the Depot in 1998 to put their views about the future to the developers. It was a week-end of consultation and discussion. The children of Caterham declared their interest, illustrating their ideas with plans and drawings. The small group behind the publication of this book arranged a photographic display showing aspects of Depot life and the many changes it underwent in peace and war.

One thought found unanimous support. The development of the Depot should retain many of its buildings, among them the Chapel (a listed building) and some of the original barrack blocks. A place should be found for them in the plans for a mixed development catering for housing, commercial purposes, leisure activities, and, perhaps a small farm.

Former Guardsmen suggested that the names of the five regiments might remain down the years in the names of some of the streets, squares and avenues. One decision has been taken.

A reminder, possibly in the form of an obelisk, will stand at the main entrance to the finished development — opposite the site of the old guardroom and on the spot where generations of sentries stood guard.

An Artists's Impression of how the Re-born Depot might look . . .

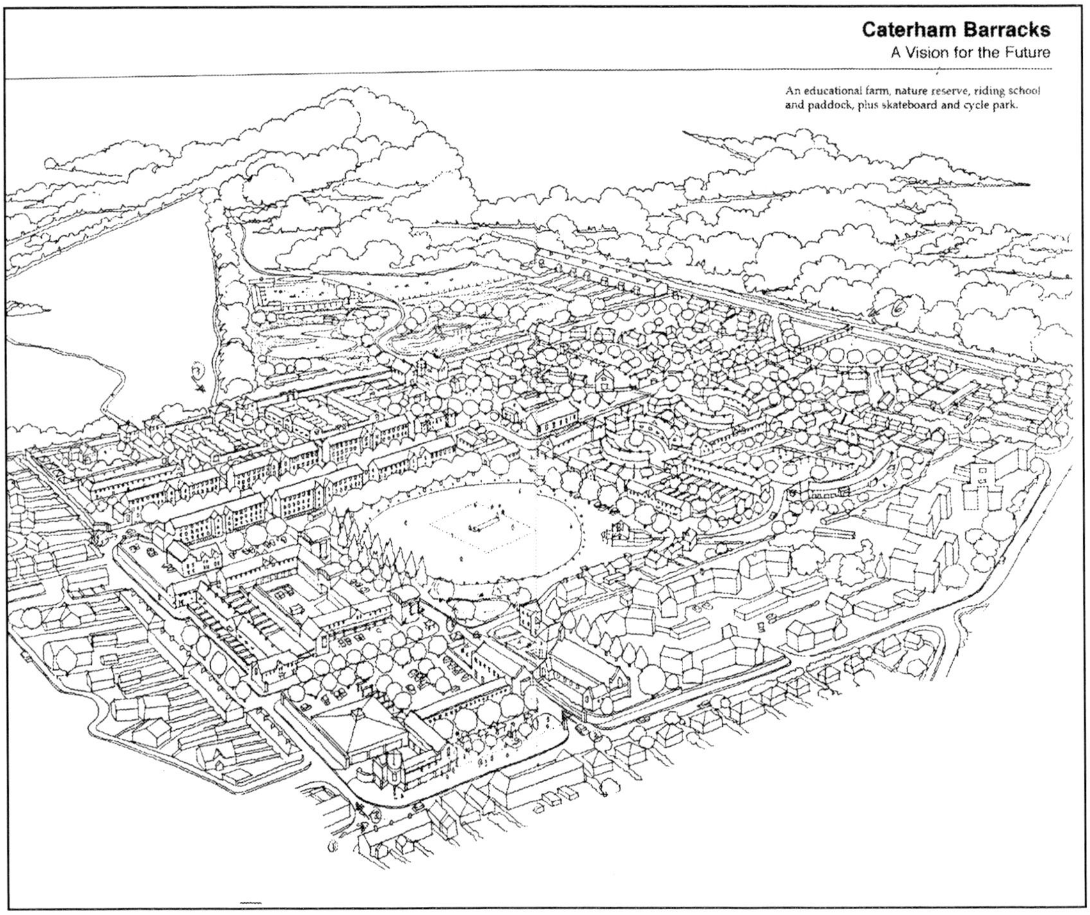

. . . with many of the Original Features retained

An Early Planning Decision

The Barrack Blocks . . .

. . . become Town Houses in a Leafy Boulevard.

The History's 'history'

The idea of a Caterham Depot history was suggested by Grenadier Archivist, the late Major Peter Lewis, MBE, in a postscript to a lecture he gave to the town's Residents' Association in April 1996. In his audience was local resident Geoffrey Hare, then involved in a history project on the Croydon Airport. He discussed the idea with former Grenadier John Tilbury of Old Coulsdon, who set about testing support for it in the Regimental Headquarters at Wellington Barracks and among the hundred-plus Associations of the five Regiments of Foot in this country and overseas.

The reaction was immediate and positive. Reminiscences about the old Depot began coming in from former Guardsmen, evidence of a strong level of enthusiasm for the project.

The next question of who would compile and edit the hook was answered by a chance encounter between Hare and Ronald Melvin, a former Scots Guardsman and ex-journalist. The meeting of Hare, Melvin and Tilbury which followed, immediately set the project on its way.

The division of tasks was automatic. Tilbury took on the communications and logistics, announcing the plans through *The Guards Magazine*, the Regimental Associations and the RHQs, and maintaining links with contributors. While Melvin researched and wrote chapters on the origins of the Depot and edited contributions, Hare employed his skills as a computer and word processor expert to put the developing manuscript on disk, on the lines of the chosen design and layout. All three were responsible for identifying and collecting illustrations for the book.

*

It is intended that sales of the history will aid Guards' charities.

In the 1800s . . .

The architect's design for the original two storey barrack blocks.

The Supporters

The organisers of this informal record of a unique establishment had the enthusiastic co-operation of former Guardsmen, many of them now thousands of miles and long years away from Caterham. Theirs is the story in these pages.

Much valuable input and encouragement came by letter and telephone from men and women with Guards connections whose names are not in the narrative. We who have been responsible for producing this book offer them our sincere thanks by naming as many as we can.

Ronald Melvin Geoff Hare John Tilbury

Supporters

Walter Gaskin,	Librarian *(retired)* and local historian, Caterham.
Roger Packham,	Chairman, Bourne *(local history)* Society, Caterham.
Richard Abbey,	St. Helier. Jersey C.I.
P. Abernethy,	Ayrshire Branch, Scots Guards Association.
D. Adams,	Ulster Branch, Irish Guards Association.
Rodney Angell-Baker,	Gomersal, West Yorkshire.
K. G. Axworthy,	Plymouth Branch, Coldstream Guards Association.
RSM. R. Appleby MBE,	Royal Hospital, Chelsea, London SW3.
O. Baddeley,	Manchester Branch, Grenadier Guards Association.
William A. Bain,	Braemar. Aberdeenshire.
Frank Ball.	
B. L. Barber,	Norfolk Branch, Grenadier Guards Association.
Capt. M. C. L. Barnett,	Equerry, RHQ. Scots Guards.
Tom Barry,	London Branch, Irish Guards Association.
G. M. Beavis,	March Branch, Grenadier Guards Association.
C. Beck,	Shropshire Branch, Grenadier Guards Association.
Ms. Gillian Bell,	Surrey Mirror.
Mrs. Olwen Bell,	Bournemouth. Dorset.
M. Berg,	Dorset and Hants. Branch, Coldstream Guards Association.
C.V.C. Booth-Jones,	Monkton Deverill. Wiltshire.
Boxer Brown.	
L. C. Bozeat, MM,	Kingston Branch, Grenadier Guards Association.
D. Brignull,	Chesterfield Branch, Grenadier Guards Association.
Andy Britton,	Hon. Secretary, Guards Association of Western Australia.

P. Brooks-Tee,	Carlisle Branch, Scots Guards Association.
Major C. Brown,	RHQ. Scots Guards.
W. J. T. Brown,	East Kent Branch, Grenadier Guards Association.
F. Bull.	
John Bushby (dec'd.),	Late Curator, East Surrey Museum. Caterham.
Michael Bushell,	Fulham, London SW6
Alan Candlish,	Hon. Secretary, Guards Association of New South Wales, Australia.
D. W. Coles,	Manchester Branch, Coldstream Guards Association.
P. L. Cook,	Oxfordshire Branch, Grenadier Guards Association.
L. Corns,	Wigan & St. Helens Branch, Grenadier Guards Association.
P. Cosh,	Weston-on-the-Green. Oxon.
P. J. Coughlan,	Leicester Branch, Coldstream Guards Association.
G. Coulson,	Windsor Branch, Coldstream Guards Association.
K. Coursey,	Regimental Archivist, Scots Guards Association.
E. Costello,	Doncaster Branch, Coldstream Guards Association.
N. A. Crockett,	Fife Branch, Scots Guards Association.
Don Cross,	The Editor, Guardsmans Post, Australia.
Henry A. Curry,	Morwell, Victoria, Australia.
G. Davidson,	Highlands, Islands & Moray Branch, Scots Guards Association.
Bill Davies,	Llanelli Branch, Welsh Guards Association.
Capt. D. M. Davies,	The Editor, Welsh Guards Association.
Peter Davies.	
S. E. Davis,	Watford Branch, Coldstream Guards Association.
Capt. J. E. Dean,	London Branch, Coldstream Guards Association.
Major S.H. Dehnel MBE,	Alexander Barracks, Pirbright, Surrey.
C. A. Dearing,	Suffolk Branch, Grenadier Guards Association.
R. Deasley,	West Midlands Branch, Irish Guards Association.
L. P. Deering,	Aylesbury & District Branch, The Grenadier Guards Association.
M. Dodds,	York Branch, Coldstream Guards Association.
F. W. Dooley,	Kings Lynn Branch, Coldstream Guards Association.
Captain. B. D. Double,	General Secretary, Grenadier Guards Association.
S. C. Dowland, BEM,	Surrey & East Hampshire Branch, Grenadier Guards Association.
M. Doyle,	Walsall Branch, Grenadier Guards Association.
M. Draper,	Cambridge Branch, Grenadier Guards Association.
John Elliott,	Ashington. Northumberland
H.E. Ellis,	Penyllergaer. Swansea.
John England,	Bere Regis, Dorset
O.A.Evans,	Bridgend, Mid. Glamorgan.
Robert Evans,	Director of Planning, Tandridge District Council.
'13' Evans,	RHQ. Welsh Guards.
George Fairbanks,	Letham. Perth.
Scott Farley,	Reporter, Croydon Advertiser.
Ms. Renée Faulkner,	Hon. Secretary, Guards Association of Victoria, Australia.
J. Ferguson,	Middlesborough Branch, Coldstream Guards Association.
R. Ferguson,	Newcastle Branch, Coldstream Guards Association.
Bill Fitness,	Swanley, Kent
Major R. E. Fletcher,	Regimental Archivist, RHQ. Welsh Guards.

Mrs. Gwyneth Fookes,	President, Bourne Society.
Malcolm Found,	London Branch, Grenadier Guards Association.
Derek Franks,	Walsall, West Midlands.
Gaz Gallacher BEM,	West Bridgeford, Notts
Ray Glasspell,	The Guards Museum
Gdsn Kevin Gorman,	RHQ Scots Guards
Derek Gough,	Widford, Herts
W. Green,	Walton-on-the-Naze, Essex.
W. Greenwood,	Bradford Branch, Coldstream Guards Association.
Doug. Hague,	North of England Branch, Welsh Guards Association.
John Hammond,	Coalpit Heath, Bristol
B. J. Harding,	Hon. Treasurer, North American Branch, Guards Association.
Major A. J. R. Harrison DL,	Wreay, Carlisle, Cumbria.
Dennis. Haydock,	York Branch, Coldstream Guards Association.
A. E. Hayter,	Berks. & Bucks. Branch, Scots Guards Association.
Jeff Heenan,	Caterham, Surrey.
Donald Hempstock,	Trowbridge. Wiltshire.
Ron Herrington,	Mandurah, Western Australia.
Gordon Hill,	Ledbury, Herefordshire.
A. T. Higgs,	Northants. Branch, Grenadier Guards Association.
Ray Hill,	Midlands Branch, Welsh Guards Association.
H.W.Hilton,	.Mundesley. Norfolk
F.J. "Mick" Holder,	Little Paxton. Huntingdon.
Major M. B. Holland,	Windsor Branch, Grenadier Guards Association.
J. G. Hook,	Bath Branch. Coldstream Guards Association.
Capt. D. D. Horn MISM,	The Guards Museum.
E. Howard,	Preston Branch, Scots Guards Association.
J. Howell Hughes,	North Wales Branch, Welsh Guards Association.
John Hughes,	Cardigan Branch, Welsh Guards Association.
J. Hyland,	North of England Branch, Irish Guards Association.
R. N. Jarvis,	Sutton-in-Ashfield, Nottinghamshire.
G. Juby,	Norwich Branch, Coldstream Guards Association.
Christopher Keeling,	London E1
Brian Keene,	Merthyr Tydfil Branch, Welsh Guards Association.
Mr. Dennis Kennedy,	Worksop, Notts.
T. Kennedy,	Kidderminster. Worcestershire
R. G. King,	Norwich, Norfolk
Mrs. Joan Kimberley,	Essex Branch, Grenadier Guards Association.
Derek Kirby (dec'd.)	Morley, West Australia
R. Kirk,	Grantham, Lincs.
Colin J. Knight,	Cheltenham & District Branch, Grenadier Guards Association.
G. A. Lamb,	Ashington Branch, Coldstream Guards Association.
S.A.S. Lauder,	Reigate, Surrey.
Ron Lewis,	Bristol Branch, Coldstream Guards Association.
Roy Lewis,	General Secretary, Welsh Guards Association.
G. Lindo,	Leicester Branch, Grenadier Guards Association.

Col. OJM Lindsay CBE FRHistS, Editor – The Guards Magazine

L. R. Linford,	Yorkshire Branch, Grenadier Guards Association.
A. F. Lock,	Bristol Branch, Grenadier Guards Association.
Colin Lord,	Hon. Secretary, Guards Association of Queensland.
Major C. J. Louch,	Regimental Archivist, RHQ. Coldstream Guards.
Rosemarie Lownds.	
L. MacBean,	Manchester Branch, Scots Guards Association.
C. McDonald,	Letchworth. Hertfordshire
R. McHenry,	Merseyside Branch, Scots Guards Association.
G. Mackender,	Suffolk Branch, Coldstream Guards Association.
Joe McKenna,	Hon. Secretary, Guards Association of New Zealand.
J. D. Marshall,	Lincoln Branch, Grenadier Guards Association.
A. E. Mason,	Sunderland Branch, Coldstream Guards Association.
P. Mason,	Central Scotland Branch, Scots Guards Association.
Mrs. J.M. Melvin,	Limpsfield, Surrey.
Trevor Mercy,	Hon. Secretary, Guards Association of Tasmania.
L. S. F. Mead,	Yeovil & Taunton Branch, Coldstream Guards Association.
D. Minto,	Edinburgh & The Lothians Branch, Scots Guards Association.
Derek G. Money, MBE,	Sussex Branch, Grenadier Guards Association.
Councillor Dick Moran,	Tandridge District Council.
Bill Morris,	Cardiff Branch, Welsh Guards Association.
Paul Mockett,	Aberkenfig. Mid. Glam.
Major Morgan.	
David L. Nott,	New Eltham, London
R. Oakes,	Birmingham Branch, Grenadier Guards Association.
A. A. Ormiston,	Liverpool Branch, Coldstream Guards Association.
B. Owen,	Wessex Branch, Grenadier Guards Association.
B. W. Padwick,	Reading & Oxford Branch, Coldstream Guards Association.
Miss Charlotte Parker,	Tandridge District Council.
Eric Parker.	
Captain R. J. D. Parry,	Montgomery & Shropshire Branch, Welsh Guards Association.
R. Parson-Wills,	Sussex Branch, Coldstream Guards Association.
Captain R. Paterson,	Surrey Branch, Scots Guards Association.
Len 'Tiger' Payne,	Caterham.
A. Peach,	Aldershot Branch, Coldstream Guards Association.
W. F. Pennington,	Liverpool Branch, Grenadier Guards Association.
B. Pennock,	Swansea & West Glamorgan Branch, Welsh Guards Association.
P. F. Percy,	Hon. Secretary, Coldstream Guards Association of Canada.
R. C. Pickering (dec'd),	Hon. Secretary, North American Branch. Guards Association.
Mrs. Betty Pickering,	Hon. Secretary, North American Branch. Grenadier Guards Association.
J. Pipkin, MBE,	Shropshire Branch, Coldstream Guards Association.
Mrs. Marion Pleydell,	East Surrey Museum. Caterham.
W. Potts,	Northumbria Branch, Grenadier Guards Association.
Lou Pownall,	Hon. Secretary, Guards Association of New Zealand.
T. Protheroe,	East Glamorgan Branch, Welsh Guards Association.
J.F.Pugh,	Oxted, Surrey.
John Raffaelli,	Malvern. Worcs.,
H. Rae,	Aberdeen & NE Scotland Branch, Scots Guards Association.

Frank Rein,	Bethanna Waters, Queensland, Australia.
D. Richards,	Ogmore Branch, Welsh Guards Association.
W. Ritchie,	Glasgow Branch, Scots Guards Association.
G. Roberts,	Leeds Branch, Coldstream Guards Association.
C. Robertson,	Dundee & Angus Branch, Scots Guards Association.
M. S. Robins,	North Staffordshire Branch, Grenadier Guards Association.
F. Robinson,	Southern Ireland Branch, Irish Guards Association.
P. Roden,	Sheffield Branch, Coldstream Guards Association.
Ray Rogers,	Old Coulsdon, Surrey.
S. Ross,	North Downs Branch, Irish Guards Association.
Gerald Rowe,	Hastings, East Sussex.
E. Salmon,	Derby Branch, Grenadier Guards Association.
J. Sayer,	Peterborough Branch, Coldstream Guards Association.
Reg. Seale,	Worcester & Hereford Branch, Grenadier Guards Association.
D. Sergeant,	Ashton, Preston, Lancashire.
G. Severn,	Nottinghamshire Branch, Grenadier Guards Association.
D. W. R. Shadbolt,	Matlock Branch, Grenadier Guards Association.
H. A. Shadrack, MBE,	Essex Branch, Coldstream Guards Association.
D. H. Simmons,	Caterham Branch, Grenadier Guards Association.
David Sloman,	North America Branch, Welsh Guards Association.
C. L. Smy, MBE. DCM,	Ipswich, Suffolk.
Arthur Smith,	Bury St. Edmunds, Suffolk.
WOI B. H. M. Smith MBE,	Coldstream Guards, HQ Household Division.
E. D. Smith,	Tunbridge Wells Branch, Grenadier Guards Association.
Mrs. June Smith,	Caterham, Surrey.
J. Southern,	Reading Branch, Grenadier Guards Association.
Mr. Stevenson,	Spalding Branch, Grenadier Guards Association.
John Stewart,	Soper Hall, Caterham, Surrey.
Ian Wright Stuart, BA. Dip. Arch. FSA (Scot),	Budleigh Salterton, Devonshire.
G. J. Stubbs,	Matlock, Derbyshire.
F. L. Sylvester, MBE,	Exeter Branch, Coldstream Guards Association.
John Tidridge,	Edmonton. Canada.
Dave Tilley,	Bend. Oregon.USA.
J. V. Thomas,	South Western Branch, Scots Guards Association.
Captain I. Townley,	North East Lancashire Branch, Coldstream Guards Association.
R. L. Tucker,	Bath Branch, Grenadier Guards Association.
E. R. Turner,	Gloucester Branch, Coldstream Guards Association.
H. Turner.	
Peter Wade,	Hon. Secretary, Guards Association of South Australia.
S. N. Wainwright,	Birmingham Branch, Coldstream Guards Association.
Mrs. Jan Ward,	Tandridge District Council.
K. Wardle,	Hon. Secretary, Grenadier Guards Association of South Australia.
Trevor Warr,	Pembroke Branch, Welsh Guards Association.
Mrs. Lavinia Watson,	Horsham, Sussex.
F. Webb,	Border Branch, Scots Guards Association.
S. Webb,	Durham & Yorkshire Branch, Scots Guards Association.
A. J. West, BEM,	Medway Branch, Grenadier Guards Association.

J. Westwell, BEM,	Burton-on-Trent Branch, Grenadier Guards Association.
Fred Whittle,	Levens, Cumbria.
Major R. E. Whyte,	Hon. Secretary, Scots Guards Association.
Keith William,	South Coney, Glamorgan.
M. Williams,	Aberdare Branch, Welsh Guards Association.
J. A. Wilkinson,	Burnley & Blackburn Branch, Grenadier Guards Association.
D. Witts,	Huddersfield Branch, Coldstream Guards Association.
K. Wood,	Coulsdon, Surrey.
Alan Woodhouse,	Monmouthshire Branch, Welsh Guards Association.
E. A. Woodmansey,	Hull Branch, Coldstream Guards Association.
F. G. Wrisberg,	London Branch, Scots Guards Association.
Harry Westgarth,	Anlaby Park. Hull
Cyril Williams,	Port Talbot. West Glamorgan

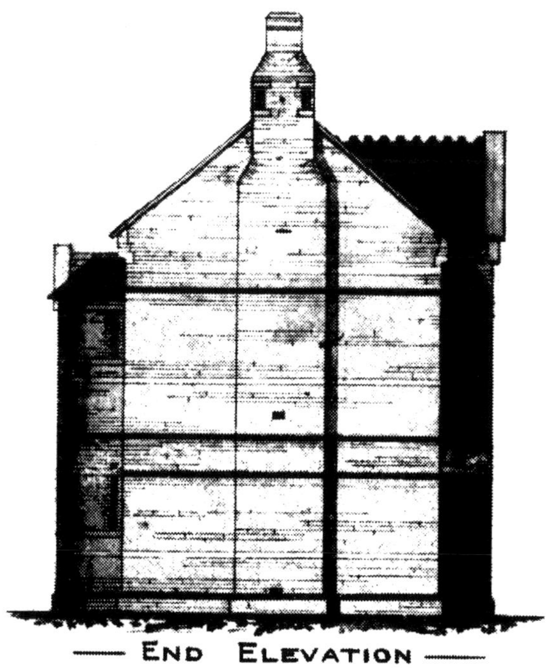

From the architect's drawing of the original two-storey barrack blocks

CYCLES - TOYS and PRAMS

I. R. RITSON

36 Westway Caterham

MISS B. E. KING *CATERHAM 2644*

Agents for Camp Surgical Supports and Berlei Brassieres and Foundation Garments

Charlotte
Corsetierre

Corset Repairs Alterations and made to measure

24 HIGH STREET, CATERHAM

W. J. SNOW

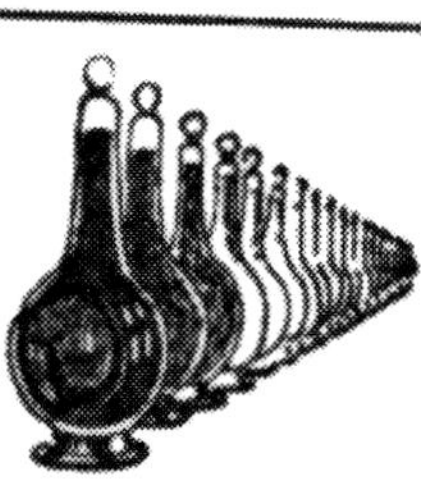

DISPENSING CHEMIST

TOILET REQUISITES
REVLON, COTY, YARDLEY, MAX FACTOR ETC.
DEVELOPING & PRINTING
KODAK ILFORD

4b High Street, Caterham-on-the-Hill

Phone 3305

Phone: CATERHAM 634

E. BELL

CYCLE, PRAM, SPORTS AND TOY STORES

32/36 WESTWAY, CATERHAM

ALSO AT

81 STREATHAM HILL, S.W.2. *Phone: TULSE HILL 2955*

10 CROWN PARADE S.E.19.

AGENT FOR :—

ALL LEADING CYCLE & PERAMBULATOR MANUFACTURERS

F. EDWARDS

Est. 50 YEARS

Fruiterer and Greengrocer, etc,

Orders taken for Wreaths, Crosses, Bouquets and all Floral Designs—Executed by expert workmen—not on premises

ORDERS DELIVERED IN THE DISTRICT

OPEN 8 a.m. to 5-30 p.m.
CLOSED 1 p.m. to 2 p.m
EARLY CLOSING—WEDNESDAY 1 p.m.

180 COULSDON ROAD, CATERHAM

Published by Guardroom Publications, 33 Inwood Ave., Old Coulsdon, Surrey CR5 1LP.
Printed by RPM Reprographics, Chichester, West Sussex PO19 2PR.